Collins easy le

Maths

Ages 7–9

Sarah-Anne Fernandes

How to use this book

This book is for parents who want to work with their child at home to support and practise what is happening at school.

- Ask your child what maths they are doing at school and choose an appropriate topic. Tackle one topic at a time.
- Help with reading the instructions where necessary, and ensure that your child understands what to do.
- Help and encourage your child to check their own answers as they complete each activity. Discuss with your child what they have learnt.
- Let your child return to their favourite pages once they have been completed, to play the games and talk about the activities.
- Reward your child with plenty of praise and encouragement.

Special features

- **Games:** There is a game on each double page that reinforces the maths topic. Each game is for two players, unless otherwise indicated. Some of the games require a spinner, which is easily made using the circles printed on the games pages, a pencil and a paper clip. Gently flick the paper clip with your finger to make it spin.

- At the bottom of every page you will find footnotes that are **Parent's notes**. These are divided into '**What you need to know**', which explain the key maths idea, and '**Taking it further**', which suggest activities and encourage discussion with your child about what they have learnt. The words in bold are key words that you should focus on when talking to your child.

ACKNOWLEDGEMENTS
The author and publisher are grateful to the copyright holders for permission to use quoted materials and images.
p.4 © erwin cartoon/Shutterstock.com; p.6 © Xtremest/Shutterstock.com; p.6 © katalina/Shutterstock.com; p.19 © Incomible/Shutterstock.com; p.22 © mystel/Shutterstock.com; p.22 © TashaNatasha/Shutterstock.com; p.26 © wongstock/Shutterstock.com; p.27 © blambca/Shutterstock.com

Every effort has been made to trace copyright holders and obtain their permission for the use of copyright material. The author and publisher will gladly receive information enabling them to rectify any error or omission in subsequent editions. All facts are correct at time of going to press.

Published by Collins
An imprint of HarperCollins*Publishers*
1 London Bridge Street
London SE1 9GF

© HarperCollins*Publishers* Limited
ISBN 9780007559817
First published 2014
10 9 8 7 6 5 4 3 2

All rights reserved. No part of this publication may be reproduced, stored in a retrieval system, or transmitted, in any form or by any means, electronic, mechanical, photocopying, recording or otherwise, without the prior permission of Collins.

British Library Cataloguing in Publication Data.
A CIP record of this book is available from the British Library.

Publishing Manager: Rebecca Skinner
Author: Sarah-Anne Fernandes (SolveMaths Ltd)
Assistant author: Gareth Fernandes (SolveMaths Ltd)
Commissioning and series editor: Charlotte Christensen
Project editor and manager: David Mantovani
Cover design: Susi Martin and Paul Oates
Inside concept design: Lodestone Publishing Limited and Paul Oates

Text design and layout: Q2A Media Services Pvt. Ltd
Artwork: Rachel Annie Bridgen, Q2A Media Services
Production: Robert Smith
Printed and bound by Printing Express Limited, Hong Kong

5 EASY WAYS TO ORDER
1. Available from www.collins.co.uk
2. Fax your order to 01484 665736
3. Phone us on 0844 576 8126
4. Email us at education@harpercollins.co.uk
5. Post your order to: Collins Education, FREEPOST RTKB-SGZT-ZYJL, Honley, HD9 6QZ

Contents

Numbers to 1000	4
Number patterns	6
Addition	8
Subtraction	10
Multiplication and division facts	12
More multiplication and division	14
Fractions	16
More fractions	18
Measures	20
Money	22
Perimeter	24
Time	26
Shape	28
Graphs	30
Answers	32

Numbers to 1000

Place value

- Write the value of the digit **7** in each of these numbers:

Comparing and ordering

- Write these numbers and measures in the correct order.

| 176 | 175 | 234 | 98 | 252 |

Most ─────────────────────────── Least

| 742 g | 756 g | 829 g | 824 g | 731 g |

Heaviest ─────────────────────── Lightest

| 185 cm | 163 cm | 165 cm | 180 cm | 167 cm |

Tallest ─────────────────────── Shortest

What you need to know At this stage your child is learning to be able to read and write **numbers to at least 1000** in **numerals** and **words**. They are also learning to recognise the **place value** of each **digit** in a **3-digit number**. For example, 236 = 200 + 30 + 6 means that the number 236 is made up of 200 (2 **hundreds**), 30 (3 **tens**) and 6 (6 **units** or 6 **ones**). This knowledge should help your child to compare and order numbers. It is always best to start with the **hundreds**, then the **tens** and then the **units** (**ones**). If, however, two or more numbers in the set have the same number of hundreds or tens, then look at the value of the digit in the next **place value column**.

Game: Crooked rules

You need: a 1–6 dice, pencil, paper.

- Use the template below to make up a playing board.
- Take turns to roll the dice.
- Write the number you rolled in a place value column. You can write the number in your own row or in your opponent's row.
- Carry on until all the place value columns are filled and both players have a 3-digit number.
- The player with the smaller number wins a point.
- Play another five rounds. The player who has more points wins!

TIP: Think about your tactics in this game. If you roll the digit '1', will you put this digit in one of your place value columns or will you give it to your opponent? Remember, the aim is to make a small number!

	Hundreds	Tens	Ones
Player 1			
Player 2			

Making and writing numbers

Jolly joke
Why is 6 afraid of 7?
Because 7 ate (8) 9!

Use the digits 1, 5 and 9 to make five 3-digit numbers.

- Write each number you make in numerals and words. An example has been done for you.

159	one hundred and fifty-nine

Taking it further Use the 'Crooked rules' board again. Play with an opponent. Roll the dice three times to make a 3-digit number, then let your opponent do the same. Look at the numbers that have been made and decide whether to use <, > or = between the two scores. For example, player 1 rolls the digits 6, 2 and 3 to make the number 326. Player 2 rolls the digits 4, 1 and 5 to make the number 541. Therefore, 326 < 541. Saying the number sentence out loud can help: "326 is less than 541". Use the language **less than** (<), **more than** (>) and **equal to** (=).

Number patterns

Number sequences

- Continue each number sequence.

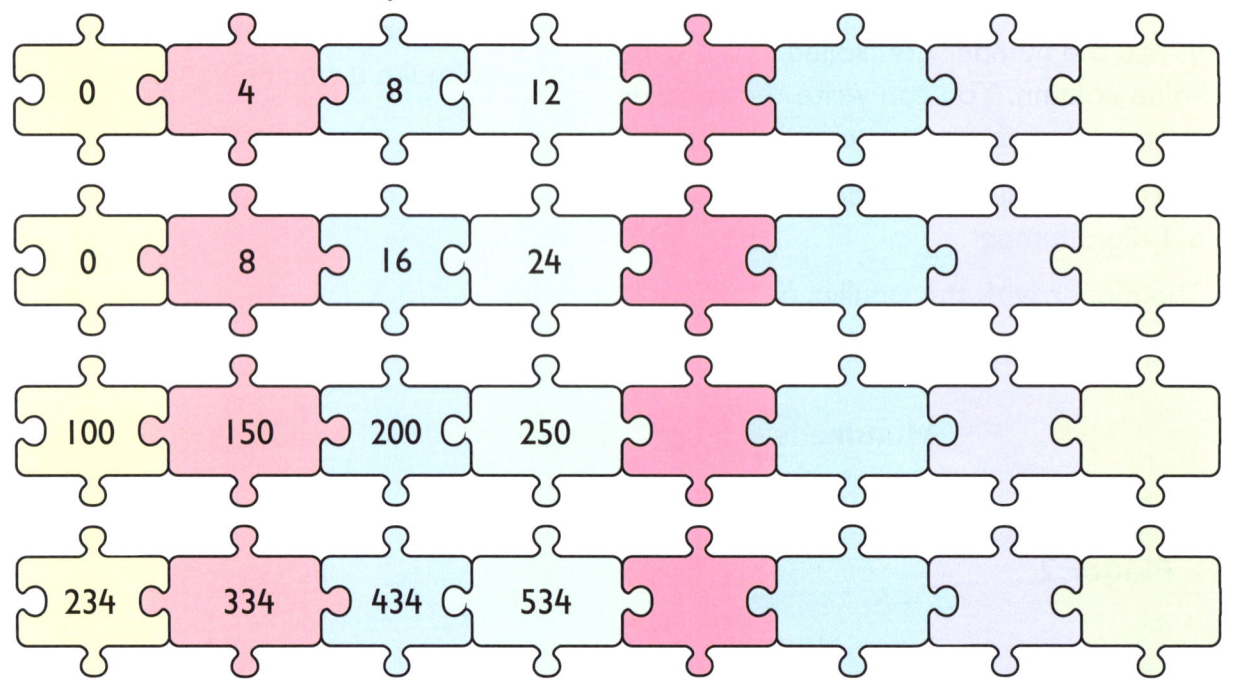

Missing numbers

- Fill in the missing numbers by finding the pattern for each sequence.

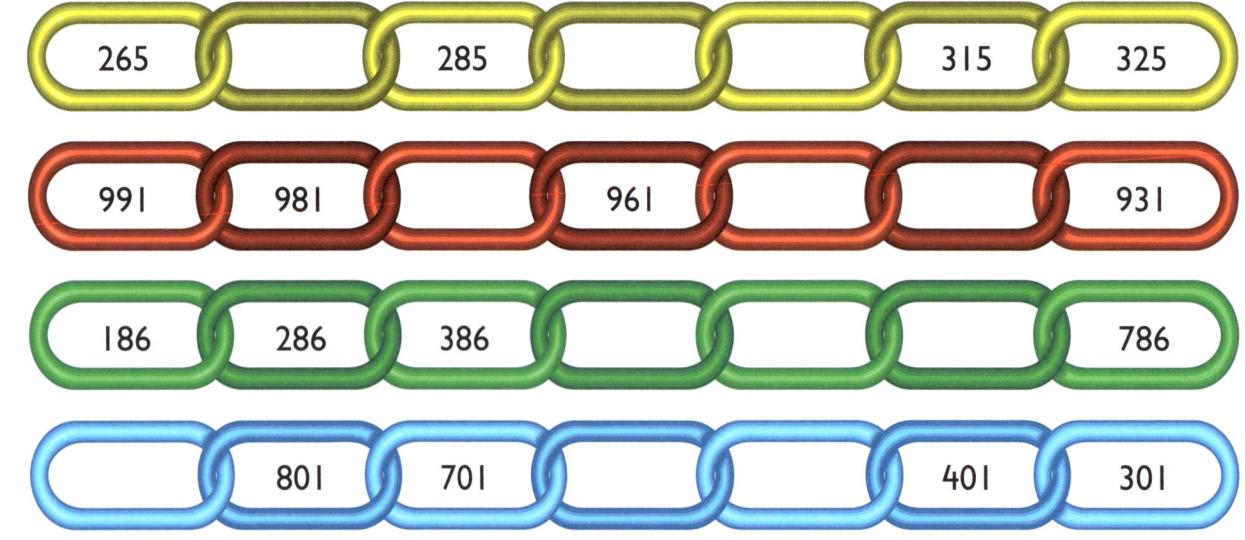

What you need to know   At this stage your child is learning to count in **multiples of 2, 4, 5, 8, 50** and **100**. They also need to know how to find **10 or 100 more** or **less** than a given number. Being able to count on and back in repeated steps will help your child with addition and subtraction. It is important to encourage your child to count on from **different multiple starting points** when counting in 4s, 8s and 50s and from **any given number** when counting in 10s and 100s, so that they are not always starting at zero.

Game: Forty-eight

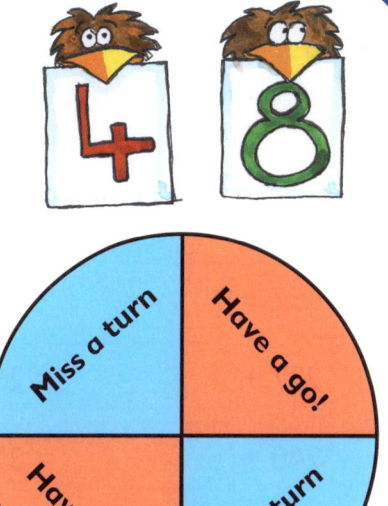

You need: a paper clip, a pencil, a coin.

- Both players must toss a coin: if you toss 'Heads' then you will be counting in eights. If you toss 'Tails' then you will be counting in fours.
- Take turns to spin the spinner. If the spinner lands on 'Have a go!' then count on one step in your step size (either eight or four) starting at 0. Jot down your number.
- On each turn, continue counting on from your last number.
- The first player to reach **48** wins a point.
- Play three rounds, beginning each time with tossing the coin. The player who has more points wins!

Cross number puzzle

Jolly joke
What did nine say to eleven?
You are out of order!

- Use number patterns to fill in the missing numbers.

			16	24		40
				34		44
35						
40		52	54			
		100	150		250	

Taking it further Extend the game 'Forty-eight' and change the target number to 80. This time rather than counting from zero choose a different multiple starting point, for example 32. Also, try the game with other multiple pairs – one player could count in 5s and the other in 10s, counting to 120 from 0.

Addition

Mental calculations

- Fill in the missing numbers.

437 + ☐ = 442   672 + ☐ = 712

237 + ☐ = 297   133 + ☐ = 433

240 + ☐ = 320   462 + ☐ = 962

Written calculations

- Calculate the answers to these additions. The first one has been done for you. You might calculate using the method shown or another written method.

136 + 147 = 283
```
          10
    100   30   6
+   100   40   7
    ‾‾‾‾‾‾‾‾‾‾‾‾
    200   80   3
```

176 + 23 = ☐

103 + 89 = ☐

341 + 75 = ☐

456 + 231 = ☐

756 + 127 = ☐

What you need to know   At this stage your child is learning to complete mental calculations such as a **3-digit number add a single digit number** (e.g. 234 + 8); a **3-digit number add a tens number** (e.g. 234 + 60) and a **3-digit number add a hundreds number** (e.g. 234 + 200). They are also learning to add numbers with up to three digits using more **formal written methods** by setting out the calculations in **columns**. Encourage your child to use their knowledge of **place value** when completing the expanded column method (as shown in the example above). This will help your child understand the value of any digits being carried.

Game: Palindromes

Palindromes
Palindromes are numbers that read the same forwards and backwards.

'737' is a palindrome number because it reads '737' forwards and '737' backwards.

'234' is **not** a palindrome, because backwards it reads '432', which is not the same.

You need: a paper clip, a pencil, paper.

- Take turns to spin the spinner three times and write down your three digits.
- Use the numbers in any order to make a 3-digit number.
- Reverse the digits to make another number.
- Add the two numbers together.
- If the number is a **palindrome number** then you win a point.
- The first player to score three points wins!

Example:
→ 4 2 1
→ 124
→ 421
→ 124 + 421 = 545

Jolly joke
Why is 2 + 2 = 5 like your left foot?
It's not right.

888

Use the digits 0, 1, 2, 3, 4, 5, 6, 7 and 8 to make two 3-digit numbers that add together to make **888**.

Example: 674 + 214 = 888

- Write all the different ways you can find to make 888.

Taking it further   Use the digits 1 to 9. Ask your child to make two 3-digit numbers to add together to make 666, 777 and 999. Encourage them to think of number pair facts to help them. For example, 9 = 9 + 0, 8 + 1, 7 + 2, 6 + 3, 5 + 4. You can then use these number pair facts to make 999 = 888 + ???

Subtraction

Mental calculations

- Fill in the missing numbers.

164 − ☐ = 158   245 − ☐ = 215

357 − ☐ = 297   623 − ☐ = 573

476 − ☐ = 176   979 − ☐ = 279

Written calculations

- Calculate the answers to these subtractions. The first one has been done for you. You might calculate using the method shown or another written method.

```
356 − 174 = 182
      200   1
      3̶0̶0̶   50    6
    − 100   70    4
      100   80    2
```

194 − 42 = ☐

274 − 69 = ☐

376 − 123 = ☐

528 − 396 = ☐

691 − 327 = ☐

What you need to know At this stage your child is learning to complete mental calculations such as a **3-digit number subtract a single digit number** (e.g. 234 − 7); a **3-digit number subtract a tens number** (e.g. 234 − 40) and a **3-digit number subtract a hundreds number** (e.g. 234 − 200). They are also learning to subtract numbers with up to three digits using more **formal written methods** by setting out the calculations in **columns**. Encourage your child to use their knowledge of **place value** when completing the expanded column method (as shown in the example above). This will help your child understand the value of any digits being 'borrowed'.

Game: Multiples of 8

You need: a paper clip, a different coloured pencil each.

- Take turns to spin the spinner twice.
- Subtract the smaller number from the larger number. If you spin the same number twice, you miss a go!
- If your answer is a multiple of 8 then colour in a star with your pencil.
- The player to colour more stars wins!

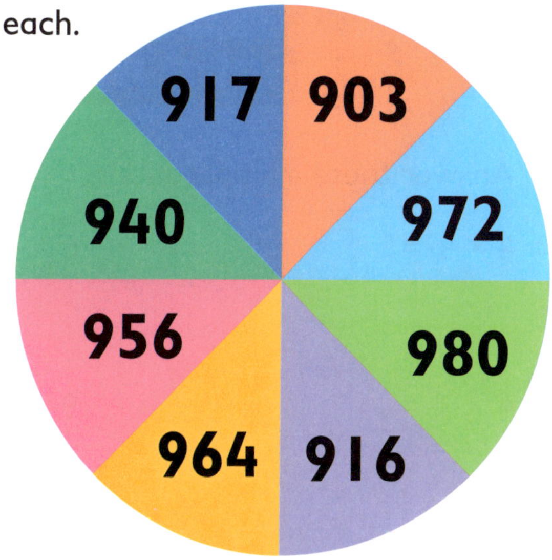

☆ ☆ ☆

☆ ☆ ☆ ☆

Jolly joke

How do you plough underground fields?

With a subtractor.

Spot the mistake

The subtractions below have some mistakes.

- Circle the mistakes in each calculation.
- Then rewrite each calculation accurately.

```
  500  40  3           700  20  9
-      20  5         - 400  70  6
  500  20  8           300  90  3
```

Taking it further Using the numbers on the spinner above, ask your child to find the difference between each of the numbers and 999. Ask your child whether they will do a formal written method or count on the difference between the two numbers. Ask your child to work out the difference between each number using their preferred strategy. Discuss how sometimes it is more efficient to use a mental calculation strategy rather than a written method.

Multiplication and division facts

Mental multiplication calculations

- Answer these multiplications.

    4 × 4 = ☐   6 × 4 = ☐   9 × 4 = ☐

    4 × 3 = ☐   9 × 3 = ☐   12 × 3 = ☐

    5 × 8 = ☐   7 × 8 = ☐   8 × 8 = ☐

   =

Mental division calculations

- Answer these divisions.

    12 ÷ 4 = ☐   28 ÷ 4 = ☐   48 ÷ 4 = ☐

    6 ÷ 3 = ☐   21 ÷ 3 = ☐   30 ÷ 3 = ☐

    32 ÷ 8 = ☐   72 ÷ 8 = ☐   96 ÷ 8 = ☐

What you need to know   At this stage your child is learning to say and use **multiplication** and **division facts** for the **2, 3, 4, 5** and **8 multiplication tables**. They will also need to learn how to **solve missing number problems** involving multiplication and division.

Game: Bubbles

You need: a paper clip, a different coloured pencil each.

- Take turns to spin the spinner.
- Multiply the number spun by 2, 3, 4, 5 or 8.
- If your answer matches a bubble then colour in the bubble using your coloured pencil.
- If the matching bubble has already been coloured, miss a go.
- The first player to colour eight bubbles wins!

Times table puzzle

Jolly joke
Teacher: "Why are you doing your multiplication on the floor?"
Child: "You told me not to use the tables."

- Find what digit each colour represents.

🟧 × 🟩 = 🟦   🟧 = ☐

3 × 🟩 = 1 🟧

🟩 × 🟩 = 1 🟨   🟦 = ☐

5 × 🟩 = 🟧 0

🟨 × 🟩 = 🟧 4

7 × 🟩 = 🟧 🟦   🟩 = ☐

🟦 × 🟩 = 3 🟧

9 × 🟩 = 3 6   🟨 = ☐

TIP: you might find it easier to start with the last multiplication fact!

Taking it further Look at the times table puzzle. Can your child write a related division fact for each multiplication fact in the puzzle?
Encourage your child to practise the multiplication tables for 2, 3, 4, 5 and 8 regularly at home. Can they spot patterns between the tables? Are there any times table songs or raps they know to help them?

More multiplication and division

Multiplying larger numbers

- Calculate these multiplications. The first one has been done for you.
  You might calculate using the method shown or another written method.

34 × 2 = 68

30 × 2 = 60
4 × 2 = 8
    68

27 × 5 =

45 × 3 =

16 × 4 =

23 × 8 =

78 × 6 =

Dividing larger numbers

- Calculate these divisions. The first one has been done for you.
  You might calculate using the method shown or another written method.

98 ÷ 2 = 49

```
    4 9
2)9 ¹8
```

65 ÷ 5 =

63 ÷ 3 =

56 ÷ 4 =

96 ÷ 8 =

84 ÷ 6 =

What you need to know At this stage your child is learning to solve more complex multiplication and division problems using their known multiplication and division facts for the **2, 3, 4, 5, 6, 8** and **10 multiplication tables**. They are learning how to multiply and divide a **2-digit number by a 1-digit number** using a written method, as shown in the examples above. They are also beginning to solve simple scaling problems, e.g. double recipe quantities or "the new building is four times as high."

Game: Divide and move

You need: a paper clip, a pencil, a different coloured counter each.

- Both players put their counters on **start**.
- Take turns to spin the spinner.
- If the number on the spinner divides into the number that the counter is on, without a remainder, then move your counter on one space.
- Continue taking turns. The first person to reach **finish** wins!

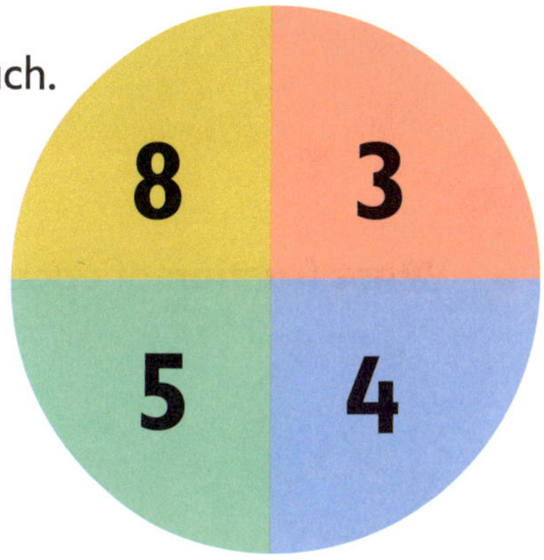

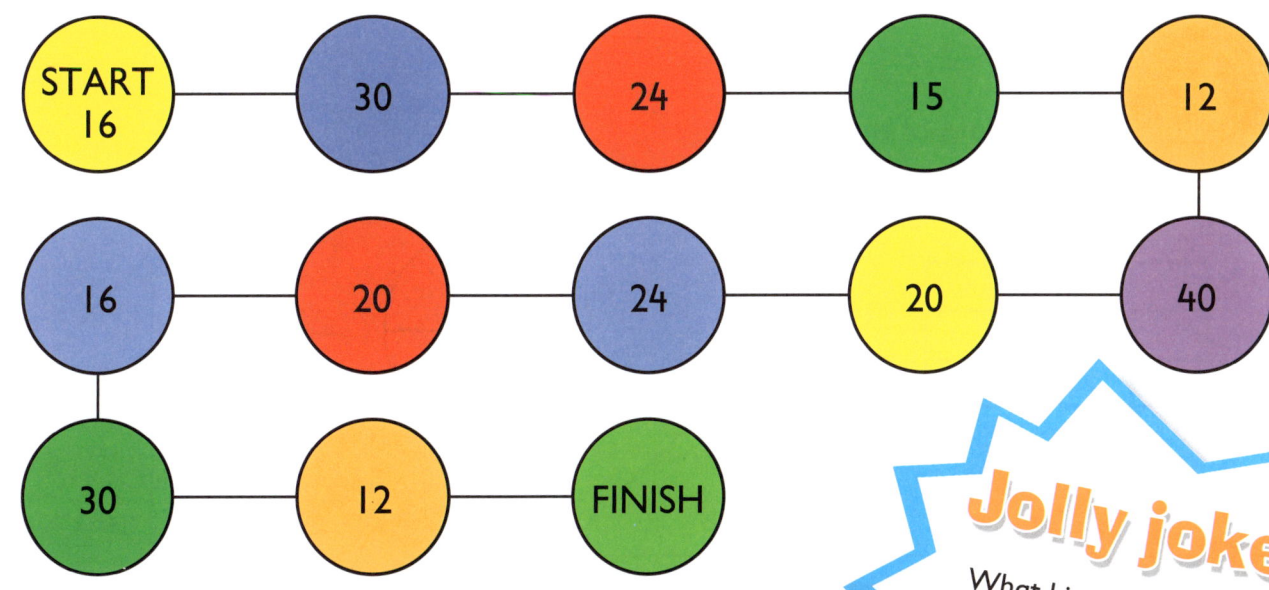

Jolly joke

What kind of pliers do you use in maths?

Multipliers!

Scaling problem

The City of London is constructing a new building that will be four times higher than the one pictured.

- Work out the height of the new building.

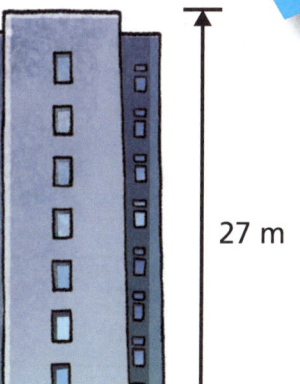

27 m

Taking it further Look at a simple recipe with your child. For example, to make a tray of flapjacks you need: 36 g of flour, 52 g of sugar, 60 g of oats, 43 g of raisins, 12 ml of honey and 32 ml of milk.
Ask your child to double the recipe, so that you have enough to share with a friend.
Then ask your child to make four times the original recipe quantity, so that there will be enough for the school fete.

Fractions

Recognising fractions

- What fraction of each shape is shaded?

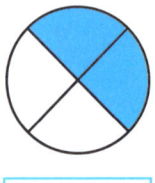 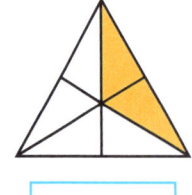 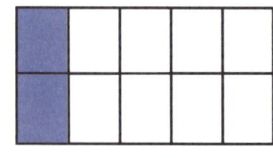 

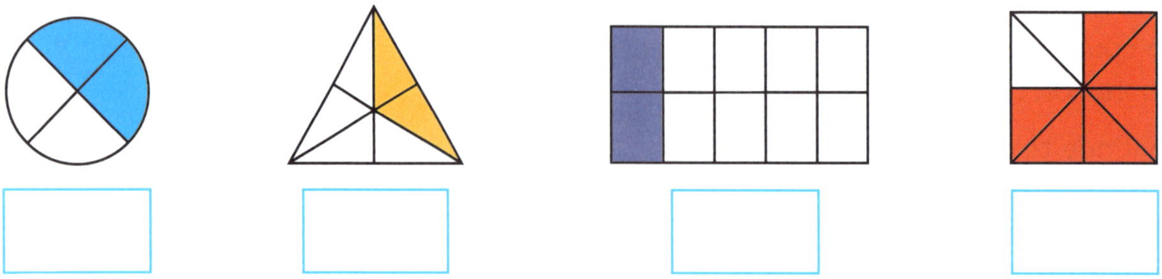

- Shade the fractions of these shapes.

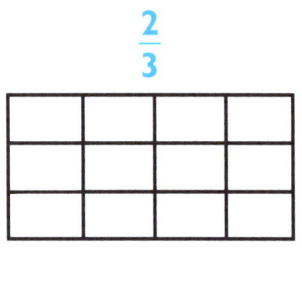 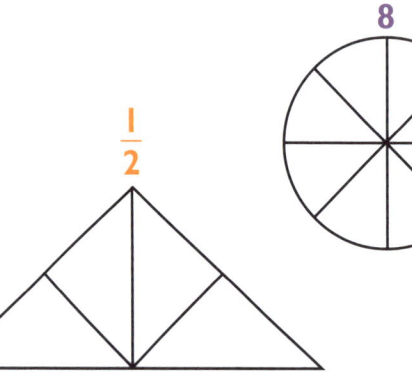 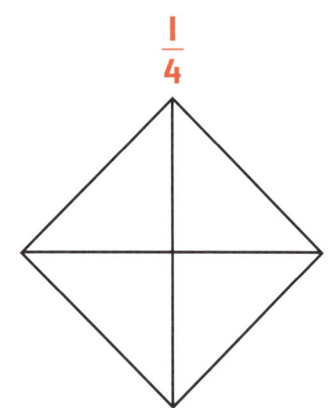

Tenths

- Complete the sequences.

$\frac{1}{10}$	$\frac{2}{10}$	$\frac{3}{10}$		$\frac{5}{10}$				$\frac{9}{10}$	$\frac{10}{10}$

$\frac{7}{10}$		$\frac{9}{10}$		$1\frac{1}{10}$	$1\frac{2}{10}$	$1\frac{3}{10}$		$1\frac{5}{10}$	

	$\frac{9}{10}$	$\frac{8}{10}$	$\frac{7}{10}$			$\frac{4}{10}$			$\frac{1}{10}$

What you need to know At this stage your child is learning to recognise **unit fractions** (i.e. fractions with a **numerator** – the top number of the fraction – of 1, such as $\frac{1}{4}, \frac{1}{3}, \frac{1}{5}$) and **non-unit fractions** with **small denominators** – the bottom number of the fraction – (such as $\frac{2}{3}, \frac{3}{4}, \frac{4}{5}, \frac{2}{6}, \frac{3}{8}$). They are learning to count up and down in tenths and learning that a tenth is when one whole object is divided into ten equal parts. They are also learning to **find a fraction of a number or amount**.

Game: Fraction maze

You need: a coloured pencil. This game is for one person only.

- Find the route through the grid from the **start** box to the **finish** box. Colour in your route with your coloured pencil.
- You can only move to a box that has a **correct calculation** in it. You can move left, right, up or down but not diagonally.

START	$\frac{1}{10}$ of 100 = 10	$\frac{2}{4}$ of 24 = 12	$\frac{2}{3}$ of 36 = 24	$\frac{2}{10}$ of 60 = 40
$\frac{1}{10}$ of 100 = 1	$\frac{1}{2}$ of 30 = 16	$\frac{1}{3}$ of 15 = 7	$\frac{2}{3}$ of 12 = 8	$\frac{1}{4}$ of 24 = 8
$\frac{2}{3}$ of 9 = 5	$\frac{2}{5}$ of 20 = 9	$\frac{1}{2}$ of 12 = 6	$\frac{1}{3}$ of 15 = 5	$\frac{1}{4}$ of 16 = 5
$\frac{2}{5}$ of 25 = 16	$\frac{1}{4}$ of 8 = 3	$\frac{2}{3}$ of 27 = 18	$\frac{1}{4}$ of 32 = 9	$\frac{2}{10}$ of 50 = 30
FINISH	$\frac{1}{4}$ of 44 = 11	$\frac{1}{3}$ of 6 = 2	$\frac{1}{10}$ of 90 = 90	$\frac{2}{3}$ of 12 = 10

Fractions of sweets

- Would you rather have $\frac{1}{3}$ of 30 sweets or $\frac{2}{5}$ of 30 sweets? Explain your choice.

Jolly joke

Who invented fractions?

Henry the eighth!

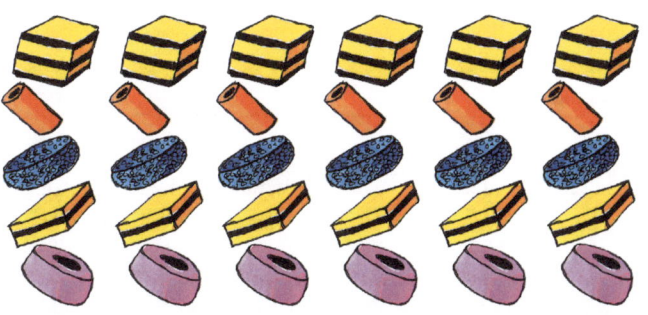

TIP: There are 30 sweets here. Use them to help you answer the question.

Taking it further Practise counting up and down in tenths from 0 to 5. Encourage your child to recognise that $\frac{5}{10}$ is the same as $\frac{1}{2}$. Continue practising finding fractions of amounts. You could ask your child to correct the various incorrect fraction calculations in the fraction maze above.

More fractions

Ordering fractions

- Write these fractions in order, from largest to smallest.
  You can use the fraction wall on page 19 to help you.

  $\frac{1}{10}$ $\frac{1}{6}$ $\frac{1}{2}$ $\frac{1}{4}$ $\frac{1}{3}$

Largest	Smallest

  $\frac{2}{8}$ $\frac{1}{8}$ $\frac{7}{8}$ $\frac{4}{8}$ $\frac{5}{8}$

Largest	Smallest

Fraction calculations

- Answer these fraction additions.

  Example: $\frac{1}{5}$ ←Add the numerators together→ $\frac{2}{5}$ $\frac{3}{5}$

  $\frac{5}{7} + \frac{1}{7} =$ ☐ $\frac{2}{6} + \frac{3}{6} =$ ☐ $\frac{1}{4} + \frac{2}{4} =$ ☐

- Answer these fraction subtractions.

  Example: $\frac{5}{6}$ ← Subtract the numerators → $\frac{3}{6}$ $\frac{2}{6}$

  $\frac{5}{6} - \frac{2}{6} =$ ☐ $\frac{4}{5} - \frac{1}{5} =$ ☐ $\frac{6}{8} - \frac{3}{8} =$ ☐

What you need to know At this stage your child is learning to **compare** and **order unit fractions** and **find fraction equivalences** using a diagram. They are also learning to **add and subtract fractions** with the **same denominator** within one whole.

Game: Ribbon length

You need: a paper clip, a pencil, paper.

- Take turns to spin the spinner and find the fraction of the length of the ribbon. Write down the length.
- After each turn, the player with the longer piece of ribbon scores a point.
- Play five rounds. The player with more points wins!

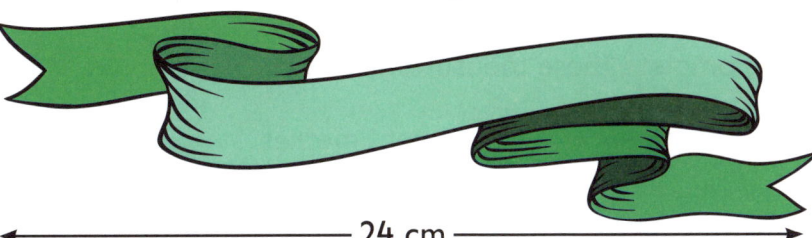

24 cm

Spinner: $\frac{1}{2}$, $\frac{3}{4}$, $\frac{1}{3}$, $\frac{1}{4}$

Jolly joke
What can a whole orange do that half an orange can't?
Look round!

Fraction equivalences

- Using the fraction wall, write the different fractions that are equivalent to $\frac{1}{2}$.

Fraction wall: 1; $\frac{1}{2}$; $\frac{1}{3}$; $\frac{1}{4}$; $\frac{1}{6}$; $\frac{1}{8}$; $\frac{1}{12}$; $\frac{1}{24}$

Taking it further Use the fraction wall to find other fraction equivalences, such as $\frac{1}{3} = \frac{2}{6}$; $\frac{1}{4} = \frac{2}{8}$; $\frac{3}{4} = \frac{18}{24}$.

Measures

Measurement facts

- Complete these measurement facts.

  1 metre = ☐ centimetres

  1 kilogram = ☐ grams

  1 litre = ☐ millilitres

  1 kilometre = ☐ metres

  4 metres = ☐ centimetres

  5 kilograms = ☐ grams

  8 litres = ☐ millilitres

  3 kilometres = ☐ metres

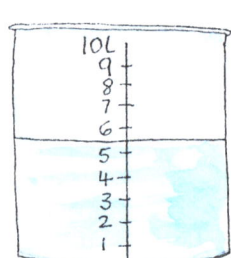

Scales

- Draw the pointer on the dial on each scale for the mass given underneath.

8 kg   **85 g**   **310 g**   **42 g**   **3 kg 500 g**

What you need to know At this stage your child is learning to choose and use appropriate measures for **length** (m/cm/mm), **mass** (kg, g) and **volume/capacity** (litre, ml). Your child is also learning how to compare, add and subtract measures.

Game: Rulers race!

You need: a paper clip, a different coloured pencil each, paper, a ruler.

- Player 1 spins the spinner and uses a ruler to draw a line equal to the length shown on the spinner.
- Player 2 does the same, but draws a separate line using their own coloured pencil.
- Player 1 spins the spinner again and now uses the ruler to extend their first line with the new length shown on the spinner. Keep a running total of the length of your line.
- Again, Player 2 repeats.
- Continue taking turns.
- The first player to draw a line more than 30 cm long wins!

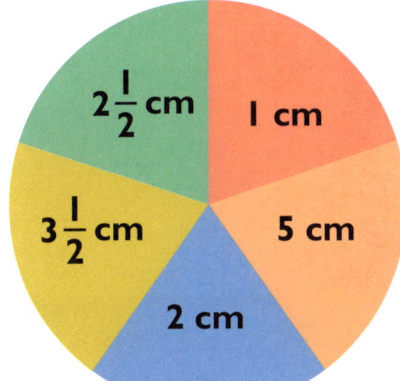

Jolly joke

What object is king of the classroom?

The ruler!

Tins of paint

- How many small tins of paint will you need to fill the large tin?

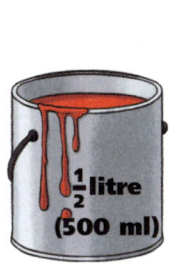

small tins

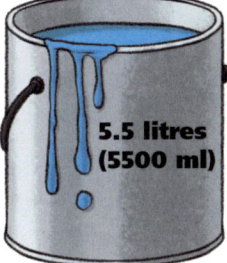

small tins

Taking it further Choose some tins, jars and packets. Ask your child to look at the mass of each item and then put them in order from heaviest to lightest.
Encourage your child to use scales at home by asking them to help measure the ingredients for a simple recipe.

Money

Finding totals

- Answer these problems by finding the totals of the price tags.

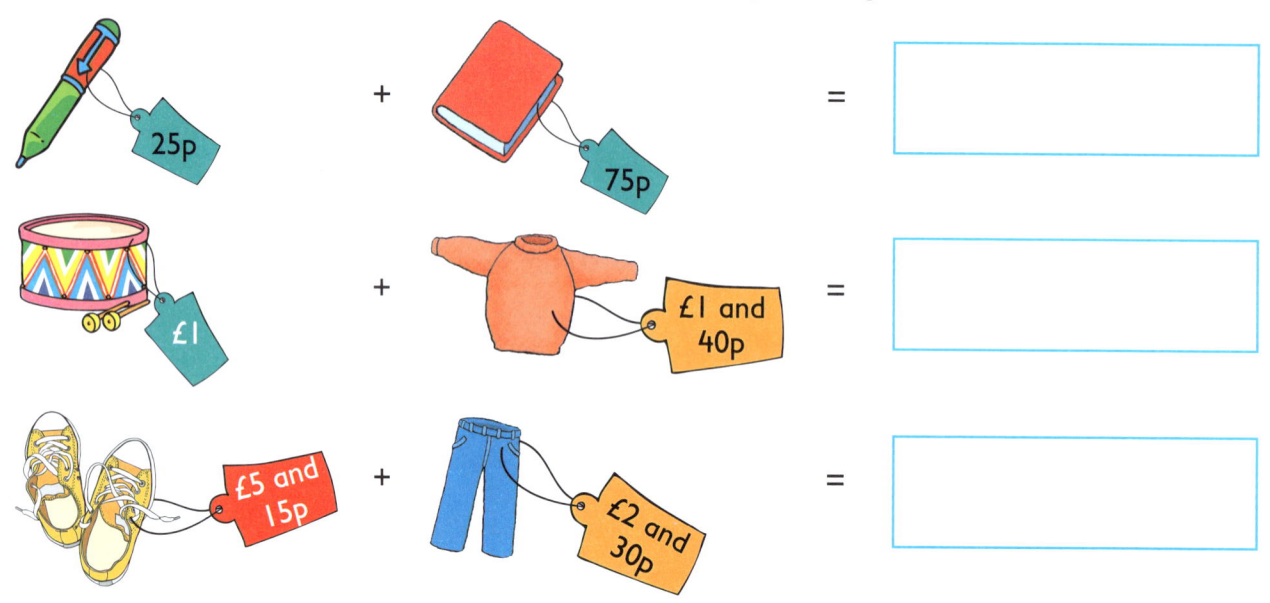

How much change?

- Work out the change for each price tag from the coin or note given.

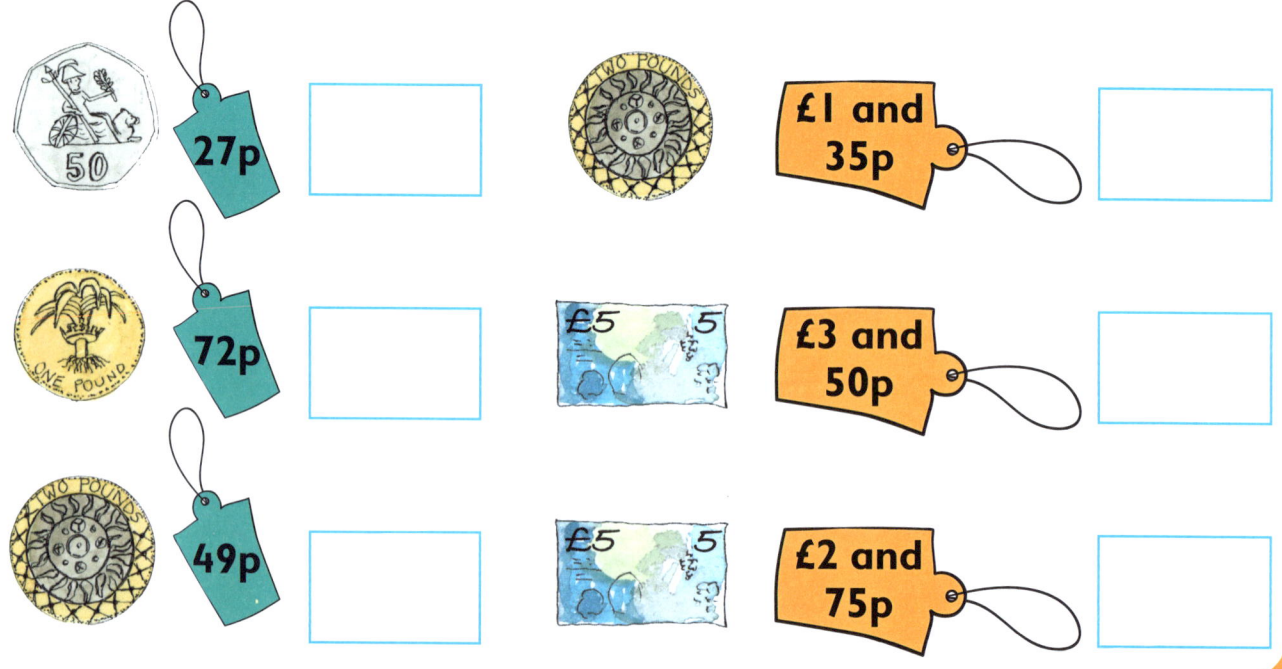

What you need to know At this stage your child is learning how to solve money problems using both **£ and p**. They need to be able to work out **total cost** and begin to find **change**. Your child needs to be confident with recognising the **value of coins** and begin to add and subtract **mixed units of money**, e.g. £1 and 20p. At this stage your child is not expected to record money using decimals.

Game: Money jar savings

You need: a 1–6 dice, a different coloured counter each, a pencil.

- Each player should choose a jar and place their counter on **start**. Take turns to roll the dice. Move your counter along the board to match the dice number rolled.
- Write the amount of money you have landed on in your money jar to show how much you have saved.
- Each player has three turns, each time saving money in the money jar.
- Add up the total saved in your money jar. The player who has saved more wins!

START →	17p	£1	74p	62p	£3	52p
						£1
28p						82p
£2		Player 1		Player 2		32p
65p	£2	63p	£1	£2	89p	£4

Jumble sale

Deka had £1 and 20p to spend. He went to a sale and bought some items. At the end of the sale he had 35p left.

- What items could he have bought?

Jolly joke

What is the quickest way to double your money?

Fold it in half!

Taking it further Play the 'Money jar savings' game again. Work out the amount of money saved in each jar. Ask your child what type of calculation they will have to do to work out how much more money they need to save to get £10. Establish that they need to find the difference between the amount saved and £10. Encourage your child to count on from the amount saved to £10 rather than doing a formal written method for subtraction.

Perimeter

Count the perimeter

- Find the perimeter of each shape by counting the squares.
  The side length of each square is 1 cm. The first one has been done for you.

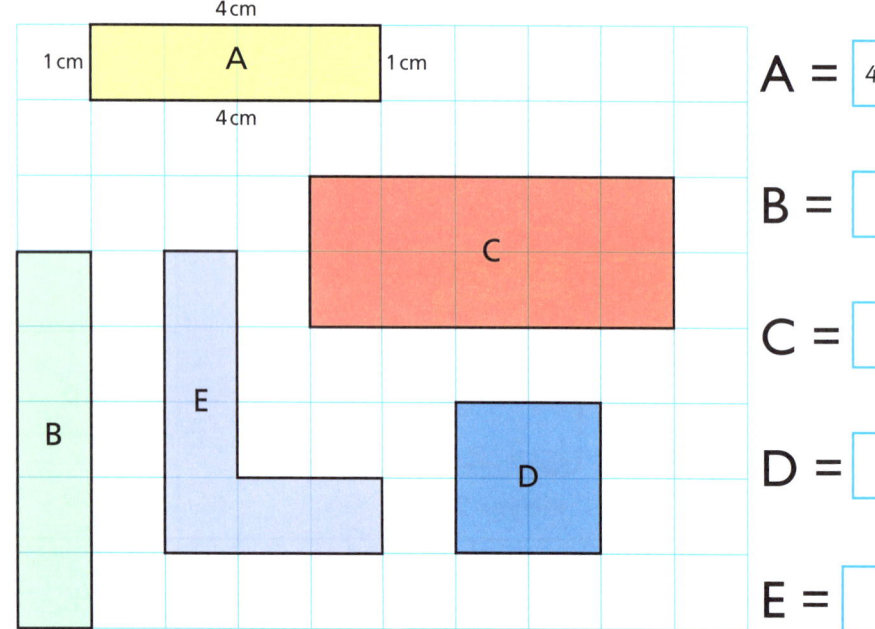

A = 4cm + 1cm + 4cm + 1cm = 10cm

B =

C =

D =

E =

Measure the perimeter

- Use a ruler to find the perimeter of each shape.

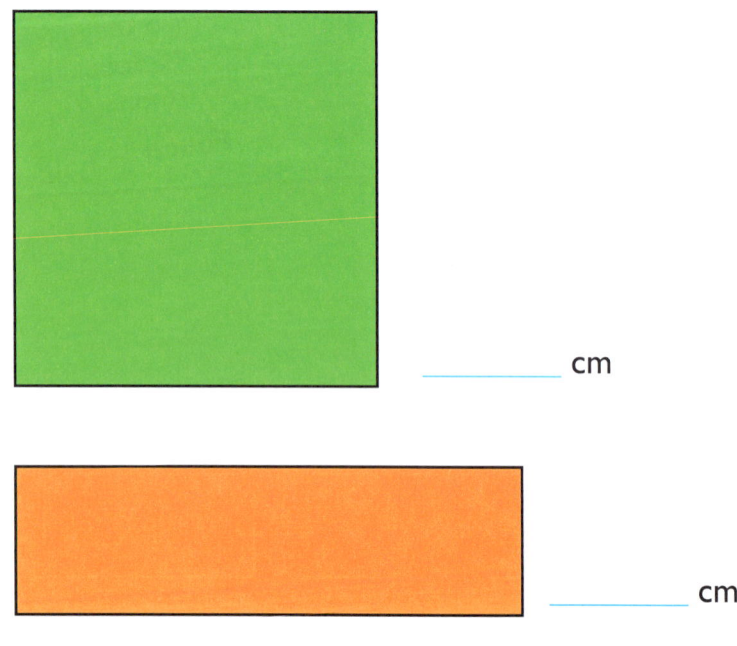

_____ cm

_____ cm

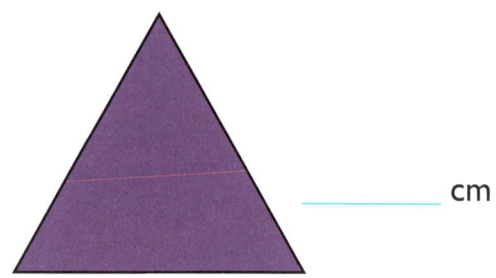

_____ cm

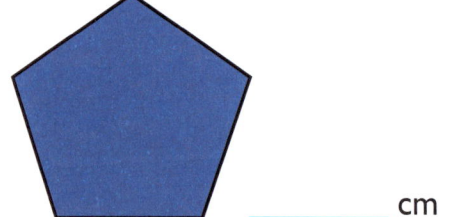

_____ cm

What you need to know At this stage your child is learning to measure the perimeter of simple 2D shapes. The perimeter is the total distance around a shape.

Game: The larger perimeter!

You need: a paper clip, a pencil, squared paper with 1 cm squares.

- Take turns to spin the spinner and look at the number.
- Draw a rectangle with that number of squares. How long is its perimeter?

Example: If you spin 6, draw a rectangle using 6 squares. Then calculate the perimeter by counting the edges of the squares. So the perimeter = 10 cm

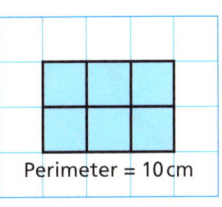
Perimeter = 10 cm

- The player with the larger perimeter wins a point.
- The first player to get five points wins!

Jolly joke
Where are teachers made?
On an assembly line!

Shapes – finding the length of one side

Write the length of one side for each of the following shapes.

- An equilateral triangle with a perimeter of 21 cm.
- A square with a perimeter of 32 cm.
- A regular pentagon with a perimeter of 50 cm.
- A regular octagon with a perimeter of 48 cm.

  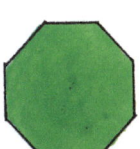 

Taking it further Ask your child to choose some square or rectangular items from around the home. Examples could be magazines or books. Ask your child to find the perimeter of each item and then order them from the smallest perimeter to the largest perimeter.

Time

Analogue and digital times

- Draw a line to match each clock face to the same time on a digital clock.

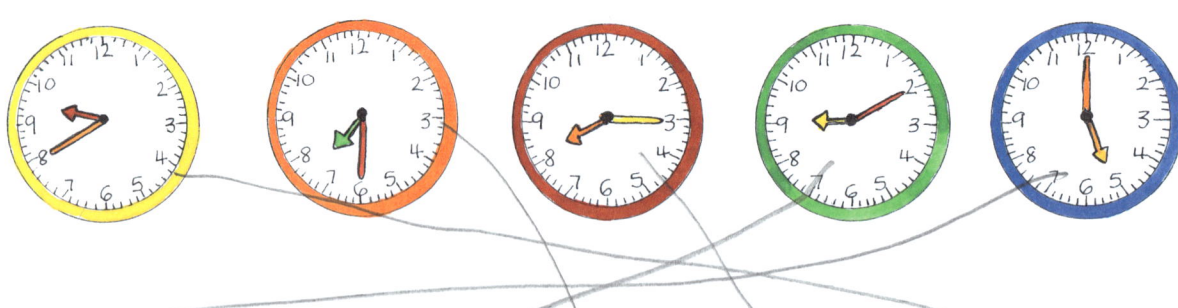

5:00   9:10   7:30   8:15   9:40

Roman numeral clock faces

Each clock face below shows a time in the morning.

- Look at each analogue clock and work out the time.
  Now write this time on the digital clock below it. The first one has been done for you.

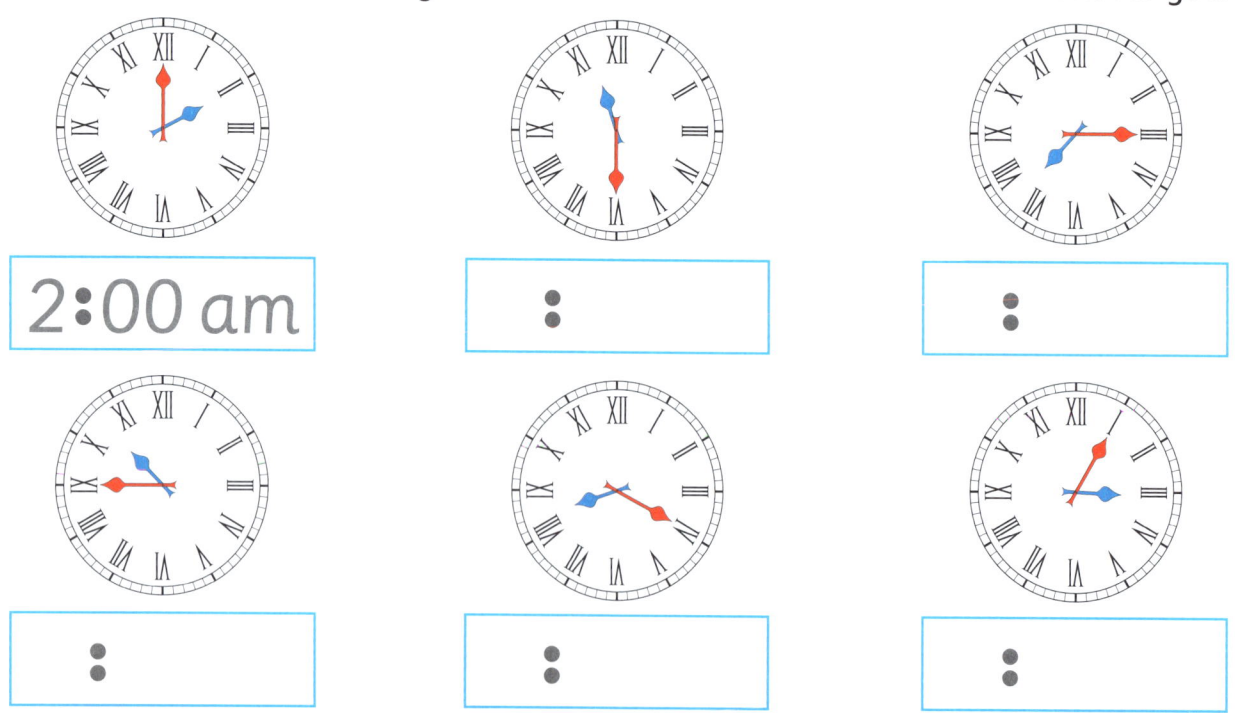

2:00 am

What you need to know At this stage your child is learning to tell and write the time from an **analogue clock**, including using **Roman numerals**. They are also learning to calculate how long a task takes.
Your child needs to learn **key time facts** such as: 60 seconds in a minute, 365 days in a year, 366 days in a leap year and the number of days in each month.

Game: Time facts

You need: a coin, a set of different coloured counters each.

- Take turns to toss the coin.
- If you toss 'Heads' then choose a yellow time question to answer. If you toss 'Tails' then choose a blue time question to answer.
- Remember to begin each question with 'How many ...?'
- If you get the question right, place one of your counters over the question. The first player to get four correct time facts in a row (vertical, horizontal or diagonal) wins!

Days in December?	Minutes in one hour?	Seconds in 2 minutes?	Days in April?	Days in a leap year?
Days in January?	Hours in 1 day?	Days in September?	Minutes in $\frac{1}{2}$ an hour?	Days in May?
Minutes in $\frac{1}{4}$ of an hour?	Days in July?	Days in March?	Seconds in 1 minute?	Hours in $\frac{1}{2}$ a day?
Hours in 2 days?	Days in 2 weeks?	Days in August?	Days in October?	Days in November?
Days in June?	Days in a week?	Minutes in $\frac{3}{4}$ of an hour?	Days in 48 hours?	Months in a year?

Bakery time problems

- Some cookies are put in the oven at 2:20 pm. They are ready at 2:50 pm. How long did they take to bake?

- Some cupcakes are put in the oven at 1:30 pm. They are ready at 2:10 pm. How long did they take to bake?

- A fruit cake is put in the oven at 3:30 pm. It is ready at 5:40 pm. How long did it take to bake?

Jolly joke
Why did the clock in the school canteen always run slow?
Every lunch it went back four seconds!

Taking it further Ask your child how long it has taken to do different activities. For example, "I put the pizza in the oven at 7:10 pm. It was ready at 7:30 pm. How long did it take to cook?" Encourage your child to use the following vocabulary accurately: am/pm, morning, afternoon, noon and midnight.

Shape

Right angles

- Draw a circle round the right angles.

2D shape drawings

- Use a ruler to draw the following 2D shapes to scale (exact size):

  1. A square: each side 4 cm
  2. A right-angled triangle: 3 cm base and 6 cm height

1

2

What you need to know At this stage your child is starting to **draw 2D shapes** and **make 3D shapes**. They are learning to recognise **angles** as a **property of a shape** and **a description of a turn**. They are also learning to **identify horizontal** and **vertical lines**, **pairs of perpendicular lines** (lines that are at right angles, 90°, to each other) and **pairs of parallel lines** (always the same distance apart and will never meet).

perpendicular   parallel

You can check right-angled triangles by using the corner of an A4 piece of paper.

Game: Identifying lines

You need: a 1–6 dice, a different coloured counter each, a pencil, paper.

- Take turns to roll the dice. Move your counter on the number of spaces from **start**.
- Look at the shape you land on and identify if the shape has parallel or perpendicular lines. Using the score card, award yourself the correct amount of points.
- Keep a running total of your points.
- The player who has more points at the end wins!

Score card	
Properties	Points
Parallel lines	10
Perpendicular lines	5
Neither	0

Rotating shapes

Sarah designs this square tile.

- She turns the tile 90°.
  Tick the tile that has the same design as Sarah's.

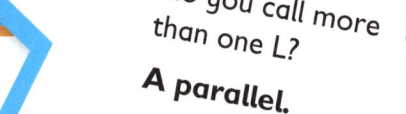

Jolly joke
What do you call more than one L?
A parallel.

Taking it further Look at a cereal box with your child. Ask what 3D shape it is. Ask them to describe the shape, taking into consideration the number of vertices (corners), edges and faces. Then unfold the box to look at how the 3D shape is made. Encourage your child to fold the box back together again to make the 3D shape.

Graphs

Pets bar chart

The bar chart shows the pets owned by children in a class.

- What is the most popular pet?
- Which pet is owned by nine children?
- How many children own a hamster?
- How many more children own a cat than a dog?
- What is the difference between the number of children who own a cat and the number of children who own a rabbit?

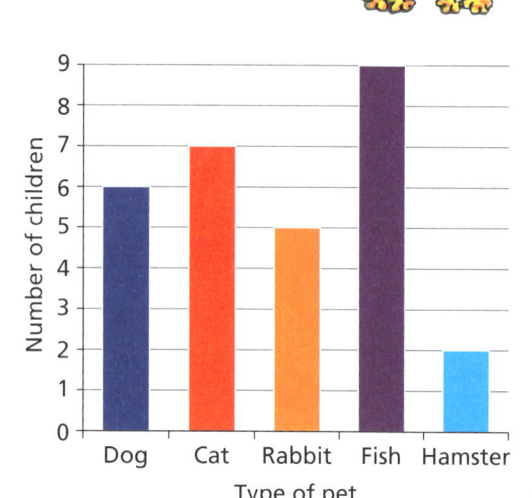

School orchestra pictogram

This pictogram shows the instruments played by children in an orchestra.

Flute	♫ ♫ ♫ ♫ ♪
Guitar	♫ ♫ ♪
Violin	
Saxophone	♫ ♪
Trombone	♫ ♪
Drums	♪
Trumpet	♪

Key
♪ = 2 students

- What is the most popular instrument in the orchestra?
- How many students play the guitar?
- How many more students play the flute than the saxophone?
- 5 students play the violin. Draw this on the pictogram.
- How many students are in the school orchestra altogether? (Don't forget about the violins!)

What you need to know At this stage your child is learning to read and draw **bar charts**, **pictograms** and **tables**. They will learn how to **solve one-step** and **two-step questions** using the information shown in the charts. Examples are "How many more …?" "How many fewer …?", and "What's the difference between …?"

Game: The best salesman

You need: a paper clip, a pencil, a ruler.

- Player 1 will sell ice cream and Player 2 will sell popcorn. Decide which player you will be.
- Player 1 spins the spinner three times and adds the numbers together.
- Player 2 spins the spinner three times and adds the numbers together.
- Each player adds their data to the bar graph.
- The player with the higher number of sales wins!

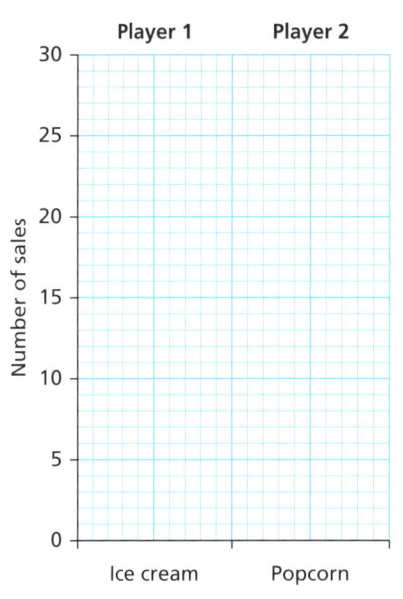

Jolly joke
What is a mathematician's favourite animal?
A grrraph!

Using data

- Use the data in the table to fill in all the missing labels on the graph.

Colour	Number of marbles
yellow	17
blue	10
red	30
rainbow	8

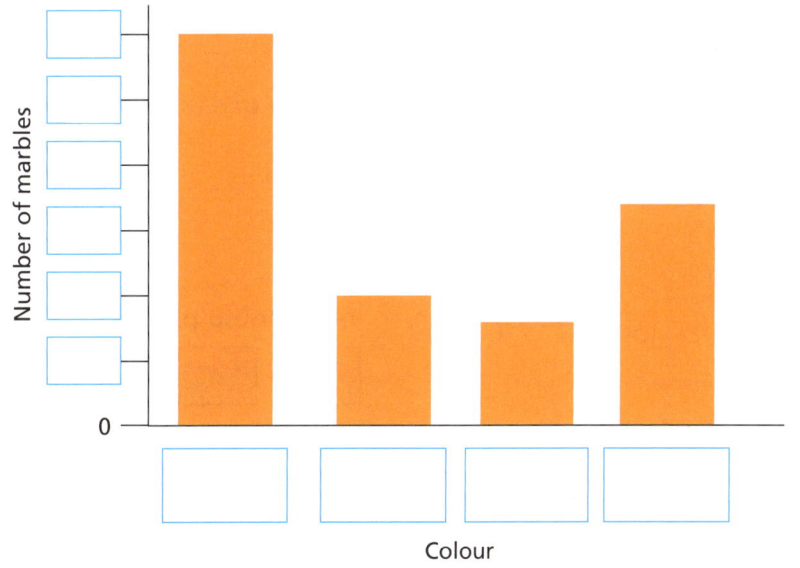

Taking it further Ask your child to carry out a survey. They could ask about their friends' favourite food, e.g. pizza, curry, burger, spaghetti. Ask your child to use this information to form a table and then draw either a bar chart or a pictogram to show the data.

Answers

Pages 4–5
Place value
70 (7 tens), 7 (7 ones/units), 700 (7 hundreds)

Comparing and ordering
252, 234, 176, 175, 98
829 g, 824 g, 756 g, 742 g, 731 g
185 cm, 180 cm, 167 cm, 165 cm, 163 cm

Making and writing numbers
195, one hundred and ninety-five
519, five hundred and nineteen
591, five hundred and ninety-one
915, nine hundred and fifteen
951, nine hundred and fifty-one

Pages 6–7
Number sequences
0, 4, 8, 12, 16, 20, 24, 28
0, 8, 16, 24, 32, 40, 48, 56
100, 150, 200, 250, 300, 350, 400, 450
234, 334, 434, 534, 634, 734, 834, 934

Missing numbers
265, 275, 285, 295, 305, 315, 325
991, 981, 971, 961, 951, 941, 931
186, 286, 386, 486, 586, 686, 786
901, 801, 701, 601, 501, 401, 301

Cross number puzzle

	8	16	24	32	40
30			34		44
35			44		48
40		52	54	56	
45					
50	100	150	200	250	300

Pages 8–9
Mental calculations
437 + 5 = 442,   672 + 40 = 712
237 + 60 = 297, 133 + 300 = 433
240 + 80 = 320, 462 + 500 = 962

Written calculations
176 + 23 = 199, 103 + 89 = 192
341 + 75 = 416, 456 + 231 = 687
756 + 127 = 883

888
Possible answers include:
102 + 786   120 + 768
123 + 765   132 + 756
201 + 687   213 + 675
234 + 654   243 + 645
312 + 576   321 + 567

Pages 10–11
Mental calculations
164 − 6 = 158, 245 − 30 = 215
357 − 60 = 297, 623 − 50 = 573
476 − 300 = 176, 979 − 700 = 279

Written calculations
194 − 42 = 152, 274 − 69 = 205
376 − 123 = 253, 528 − 396 = 132
691 − 327 = 364

Spot the mistake

$$\begin{array}{r} \cancel{5}00\ \cancel{4}0\ {}^{1}3 \\ -20\ 5 \\ \hline 500\ 10\ 8 \end{array} = 518$$

The units were added rather than subtracted. Did not carry from the tens column.

$$\begin{array}{r} \cancel{6}00\ {}^{1} \\ \cancel{7}00\ 20\ 9 \\ -400\ 70\ 6 \\ \hline 200\ 50\ 3 \end{array} = 253$$

Added the tens column rather than subtracted. Did not carry from the hundreds column.

Pages 12–13
Mental multiplication calculations
4 × 4 = 16, 6 × 4 = 24, 9 × 4 = 36
4 × 3 = 12, 9 × 3 = 27, 12 × 3 = 36
5 × 8 = 40, 7 × 8 = 56, 8 × 8 = 64

Mental division calculations
12 ÷ 4 = 3,   28 ÷ 4 = 7
48 ÷ 4 = 12,   6 ÷ 3 = 2
21 ÷ 3 = 7,   30 ÷ 3 = 10
32 ÷ 8 = 4,   72 ÷ 8 = 9
96 ÷ 8 = 12

Times table puzzle

Pages 14–15
Multiplying larger numbers
27 × 5 = 135,   45 × 3 = 135
16 × 4 = 64,   23 × 8 = 184
78 × 6 = 468

Dividing larger numbers
65 ÷ 5 = 13,   63 ÷ 3 = 21
56 ÷ 4 = 14,   96 ÷ 8 = 12
84 ÷ 6 = 14

Scaling problem
27 × 4 = 108 m

Pages 16–17
Recognising fractions
$\frac{2}{4} = \frac{1}{2}$, $\frac{2}{6} = \frac{1}{3}$, $\frac{2}{10} = \frac{1}{5}$, $\frac{6}{8} = \frac{3}{4}$

Examples are:

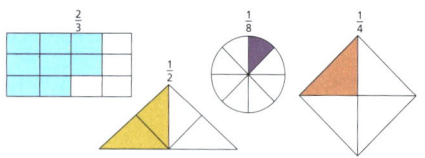

Tenths

$\frac{1}{10}$	$\frac{2}{10}$	$\frac{3}{10}$	$\frac{4}{10}$	$\frac{5}{10}$
$\frac{6}{10}$	$\frac{7}{10}$	$\frac{8}{10}$	$\frac{9}{10}$	$\frac{10}{10}$

$\frac{7}{10}$	$\frac{8}{10}$	$\frac{9}{10}$	1	$1\frac{1}{10}$
$1\frac{2}{10}$	$1\frac{3}{10}$	$1\frac{4}{10}$	$1\frac{5}{10}$	$1\frac{6}{10}$

1	$\frac{9}{10}$	$\frac{8}{10}$	$\frac{7}{10}$	$\frac{6}{10}$
$\frac{5}{10}$	$\frac{4}{10}$	$\frac{3}{10}$	$\frac{2}{10}$	$\frac{1}{10}$

Game: Fraction maze

START	✓	✓	✓	✗
✗	✗	✗	✓	✗
✗	✗	✓	✓	✗
✗	✗	✓	✗	✗
FINISH	✓	✓	✗	✗

Fractions of sweets
$\frac{1}{3}$ of 30 = 10 sweets. $\frac{2}{5}$ of 30 = 12 sweets. I would therefore prefer $\frac{2}{5}$ of 30 sweets as you get more sweets.

Pages 18–19
Ordering fractions
$\frac{1}{2}, \frac{1}{3}, \frac{1}{4}, \frac{1}{6}, \frac{1}{10}$
$\frac{7}{8}, \frac{5}{8}, \frac{4}{8}, \frac{2}{8}, \frac{1}{8}$

Complete
Drama
for Cambridge IGCSE®

Pauline Courtice
Susan Elles
Rob Thomson

Oxford excellence for Cambridge IGCSE®

OXFORD

Great Clarendon Street, Oxford, OX2 6DP, United Kingdom

Oxford University Press is a department of the University of Oxford. It furthers the University's objective of excellence in research, scholarship, and education by publishing worldwide. Oxford is a registered trade mark of Oxford University Press in the UK and in certain other countries

© Oxford University Press 2016

The moral rights of the authors have been asserted

First published in 2016

All rights reserved. No part of this publication may be reproduced, stored in a retrieval system, or transmitted, in any form or by any means, without the prior permission in writing of Oxford University Press, or as expressly permitted by law, by licence or under terms agreed with the appropriate reprographics rights organization. Enquiries concerning reproduction outside the scope of the above should be sent to the Rights Department, Oxford University Press, at the address above.

You must not circulate this work in any other form and you must impose this same condition on any acquirer

British Library Cataloguing in Publication Data
Data available

978-0-19-836674-4

10 9 8 7 6 5 4 3

Paper used in the production of this book is a natural, recyclable product made from wood grown in sustainable forests.
The manufacturing process conforms to the environmental regulations of the country of origin.

Printed in India by Multivista Global Pvt. Ltd

Acknowledgements

®IGCSE is the registered trademark of Cambridge International Examinations.

The questions, example answers, marks awarded and/or comments that appear in this book were written by the authors. In examination, the way marks would be awarded to answers like this might be different.

The publishers would like to thank the following for permissions to use their photographs:

Cover image: © Ariel Skelley/Blend Images/Corbis; p2: Clark Nobby/ArenaPAL; p3: RIA Novosti/Alamy; p7: Ted Foxx/Alamy; p6: Michael Dwyer/Alamy; p8: Malcolm Dav/ArenaPAL; p9: Suzan/ArenaPAL; p10: Hill Street Studios/Getty Images; p11: Jacques Demarthon/Getty Images; p12: Malcolm Dav/ArenaPAL; p13: Moviestore Collection/REX; p14: Malcolm Dav/ArenaPAL; p15(R): Moviestore Collection/REX; p15(L): ArenaPAL; p17: Nigel Norrington/ArenaPAL; p18: Pete Jones/ArenaPAL; p20: Robert Day/Pilot Theatre; p22: Pete Jones/ArenaPAL; p27: Photo by S Thyagarajan/Noushad Mohamed Kunju; p24: Stephen Cummiskey; p25: Courtesy of NIE Theatre; p30: Mike Abrahams/Alamy; p31: Alastair Muir/REX; p32: John F Stephenson/Getty Images; p33: Evening Standard/Getty Images; p34: Fox Photos/Getty Images; p35: Humphrey Spender/Picture Post/Getty Images; p36: Nigel Norrington/ArenaPAL; p37: Spittoon Image ©Forest Arts Centre/Company Gavin Robertson; p38: Marilyn Kingwill/ArenaPAL; p41: Mike Eddowes/Courtesy of The Nuffield Theatre; p42: Mike Eddowes/Courtesy of The Nuffield Theatre; p44: Mike Eddowes/Courtesy of The Nuffield Theatre; p46: Lewis Morley/ArenaPAL; p51: Nigel R Barklie/REX; p53: He Junchang/Xinhua Press/Corbis; p54: Nobby Clark/ArenaPAL; p55: Hindustan Time/Getty Images; p57: Pete Jones/ArenaPAL; p59: Kurt Hutton/Picture Post/Getty Images; p60: Nigel Norrington/ArenaPAL; p62: Colin Willoughby/ArenaPAL; p63(T): Malcolm Dav/ArenaPAL; p63(B): Malcolm Dav/ArenaPAL; p64: Hulton Archive/Getty Images; p65: Moviestore Collection/REX; p61: Interfoto/Alamy; p67(T): Courtesy of Laura McEwen/Pilot Theatre Company; p67(B): Courtesy of Laura McEwen/Pilot Theatre Company; p68: Malcolm Dav/ArenaPAL; p69(T): Colin Willoughby/ArenaPAL; p69(B): Alastair Muir/REX; p70: Hill Street Studios/Getty Images; p73: Lightversus/Wikipedia; p74(L): Stringer/Getty Images; p74(R): Royal Shakespeare Company; p75: Mat Hayward/Fotolia; p72: Reproduced by kind permission of Rhubarb Theatre photographer Phil Crow; p76: Robbie Jack/Corbis; p77: ImageBroker/Alamy; p78: Steppenwolf Theatre/Michael Brosilow; p79(T): DEA/A Dagli Orti/Getty Images; p79(B): Tristram Kenton/Lebrecht Music & Arts; p80: Garry Gay/Getty Images; p81: George F Mobley/National Geographic/Getty Images; p83: Nobby Clark/ArenaPAL; p84: Elliott Franks/ArenaPAL; p85: Nigel Norrington/ArenaPAL; p86: ©Icon & Co (Wales) Ltd; p87(T): Nagel Photography/Shutterstock; p87(B): Nathan Willock/Corbis; p91: Sheila Burnett/ArenaPAL; p93(T): Malcolm Dav/ArenaPAL; p93(B): Courtesy of NIE Theatre; p94: Edward Dimsdale/Filter Theatre; p97: Imaginewithme/iStockphoto; p99: Andia/ArenaPAL; p99(B): Bernie Epstein/Alamy; p100: Kai Pfaffenbach/Reuters; p101: Michael Regan/Getty Images; p99(T): Vibrant Pictures/Alamy; p103: Eddie Mulholland/REX; p104: China Photos/Getty Images; p105: Rowan Tolley/ArenaPAL; p106: Matt Cardy/Getty Images; p112: Alastair Muir/REX; p113: Jeff J Mitchell/Getty Images; p118: Robert Day/ArenaPAL; p115: Alamy; p120: Brad Barket/Getty Images; p121: Neilson Barnard/Getty Images; p122: Robert Day/ArenaPAL; p125: Robert Day/ArenaPAL; p126: Hill Street Studios/Getty Images; p131: Cineclassico/Alamy; p132: Evka119/Shutterstock.

We are grateful to the authors and publishers for use of extracts from their titles and in particular for the following:

W H Auden: "Night Mail" copyright © 1938, renewed 1966 by W. H. Auden; from W. H. AUDEN COLLECTED POEMS. Reprinted by permission of Curtis Brown Ltd and Random House, an imprint and division of Penguin Random House LLC. All rights reserved. Any third party use of this material, outside of this publication, is prohibited. Interested parties must apply directly to Penguin Random House LLC for permission.

Bertolt Brecht: *Mother Courage* © Bertolt Brecht, 2003, Bloomsbury Methuen Drama, an imprint of Bloomsbury Publishing Plc. Reprinted by permission.

Edited Twelfth Night extract courtesy of Steve Gooch, creator of 'The Cut Shakespeare' series of Shakespeare texts edited for performance and reading aloud, also the author of four dozen works for the stage including Female Transport, performed some 500 times around the world, and Writing a Play, based on his popular playwriting class.

David Foxton: 'A Memory of Lizzie' taken from *Sepia and Song* (Drama Anthologies) Nelson Thornes 2000. Reproduced by permission of David Foxton.

Franz Kafka: excerpt(s) from THE METAMORPHOSIS, IN THE PENAL COLONY, AND OTHER STORIES, translated by Willa and Edwin Muir. Published by Vintage Classics. Copyright © 1948 by Schocken Books. Copyright © renewed 1975 by Schocken Books. Used by permission of Schocken Books, an imprint of the Knopf Doubleday Publishing Group, a division of Penguin Random House LLC and The Random House Group Ltd.

The Theban Plays by Sophocles, translated with an introduction by E.F. Watling (Penguin Classics, 1947). Copyright © E.F. Watling, 1947. Reproduced by permission of Penguin Books Ltd.

Don Taylor (translator): *Sophocles Plays: One* © Don Taylor (translator), *Antigone*, Bloomsbury Methuen Drama, an imprint of Bloomsbury Publishing Plc. Reproduced by permission of Bloomsbury Publishing Plc.

Benjamin Zephaniah: 'Money (rant)' from City Psalms (Bloodaxe Books, 1992). Text copyright © Benjamin Zephaniah, 1992. Reproduced with permission of Bloodaxe Books on behalf of the author. www.bloodaxebooks.com.

Antigone copyright © Jean Anouilh and Lewis Galantière, 1951 by permission of Alan Brodie Representation Ltd. www.alanbrodie.com and Bloomsbury Methuen Drama, an imprint of Bloomsbury Publishing Plc. Original version of ANTIGONE by Jean Anouilh, Éditions de La Table Ronde, 1946.

Although we have made every effort to trace and contact all copyright holders before publication this has not been possible in all cases. If notified, the publisher will rectify any errors or omissions at the earliest opportunity.

Links to third party websites are provided by Oxford in good faith and for information only. Oxford disclaims any responsibility for the materials contained in any third party website referenced in this work.

Contents

1. Welcome to IGCSE Drama 2
2. Scripted work 8
3. Devising practical work 22
4. Design and technical work 60
5. Developing your physical skills 98
6. Preparing for a performance 112
7. Writing about your performances 120

    Glossary 134

    Index 138

Unit 1 Welcome to IGCSE Drama

Welcome! Drama is an exciting subject and one that people become very passionate about. Now that you have opened this book, come and see what we have to offer. We hope that you are already curious and are willing to have a go, and that is all we ask of you at this stage. Everything else will come to you as you work your way through the course.

Cambridge IGCSE syllabus aims

The syllabus aims to:

- develop learners' understanding of drama through practical and theoretical study
- enable learners to understand the role of actor, director and designer in creating a piece of theatre
- develop learners' acting skills, both individually and in groups
- enable learners to develop skills in devising original drama
- help learners communicate feelings and ideas to an audience
- foster understanding of the performance process and enable learners to evaluate the various states of that process
- encourage enjoyment of drama.

Did you know?
Theatre has formed an important part of many civilisations. In Ancient Greece, tragedies or comedies were performed by men or boys, often wearing masks.

1.1 Why choose drama?

Drama is a practical subject. It deals with people. Drama involves looking at people's thoughts, their feelings, and their relationships with each other and with the world about them. Through drama we can see into the lives of others and it can help us make sense of our own.

For thousands of years people have made audiences laugh or cry by acting out stories that were important to them. Even today, audiences can feel sympathy with characters created by the ancient Greeks and join in the laughter of traditional Japanese Kyōgen (comedic theatre often performed as a short intercession between longer, more solemn forms of theatre). By doing IGCSE drama you can become part of those traditions and learn how to present stories and ideas to an audience. For some, the pleasure of having a responsive, appreciative audience is the biggest reason for choosing the subject, but everyone will find something to interest and challenge them. You all will have an important part to play in whatever project you are working on.

Figure 1.1 *Japanese Kyōgen*

Drama is about working with other people

Working as part of a team is something that draws people to the study of drama. It is also a skill that attracts prospective employers, as the 'Remember' box below shows.

As you work in your group, you will learn to share ideas and to offer and take criticism constructively. Unlike many other subjects, in Cambridge IGCSE Drama you are not alone. Sometimes you will have to work under pressure to meet firm deadlines and this can be made less daunting by having the support of others in your group. This is both a comfort and a responsibility. It means that, in return, you will need to give support to others by being reliable, turning up on time for rehearsals, and by making sure that lines are learnt or that costumes or sound effects are ready on time. Throughout this book you will see that group working is stressed in every unit and this is because the success of a drama performance depends entirely on collaboration and teamwork.

Drama is about learning new skills

Drama is not just about acting. There are other important areas of skill to be learnt, involving directing and design elements. Though you may start with one area that attracts you, it is possible to branch out and try other skills. In fact, it is an advantage for a lighting designer to know what problems an actor faces on stage when designing the lighting plot, and all designers need to be aware of the practical limitations of their sets or costumes. You will learn the correct vocabulary and the methods commonly used by professionals, not by memorising lists but by *doing*.

> **Remember**
> A qualification in drama demonstrates that you have:
> - shown creativity and imagination
> - worked with others constructively
> - worked to, and met, tight deadlines
> - learnt to communicate effectively
> - interpreted your own and other people's ideas and realised them.

Welcome to IGCSE Drama

1.2 What can you expect?

> **Objectives**
>
> In this section you will learn:
>
> - how Cambridge IGCSE Drama is assessed
> - how the knowledge and skills you develop during your course are examined.

How Cambridge IGCSE Drama is assessed

The assessment areas are as follows:

- understanding of repertoire – you will be assessed on your knowledge and understanding of the possible ways of interpreting and presenting plays in the published repertoire
- devising – you will be assessed on your ability to devise and perform dramatic material, and reflect on its effectiveness
- acting – you will be assessed on your acting skills and your ability to communicate effectively to an audience.

There are two assessed components:

- Component 1 Written examination
- Component 2 Coursework

Component 1: the written examination

The written examination will be based entirely on practical work you have done, relating to pre-release material sent to your centre some months before you sit the paper. This will consist of three stimuli and an extended extract from a play. By the time you come to sit the written examination, you will have spent time studying the play extract and will have devised a piece of drama based on *one* of the given stimuli. The questions that are set require that you have engaged with the pre-release material from the point of view of actor, director and designer. In other words, the examination is focused on material you will already have experienced from a practical perspective.

The question paper will be structured in three sections:

- Section A (30 marks) consists of 6 short-answer questions on the play extract (20 marks) and 2 questions on the drama devised from your chosen stimulus (10 marks).
- Section B (25 marks) consists of one longer-answer question from a choice of three questions on the extract from a play.
- Section C (25 marks) consists of one longer-answer question from a choice of three on the drama devised from your chosen stimulus.

> **Remember**
>
> You can read about your assessments in more detail using the Cambridge IGCSE Drama syllabus at **www.cie.org.uk**.

Component 2: the coursework

You will need to submit *three* pieces of practical work, including *one* individual piece and *two* group pieces. You should have an opportunity to create more than these three pieces of work during your course so that you have a selection of pieces from which to choose your best work for submission.

The coursework components are as follows:

- one individual performance of an extract from a play, lasting between 3 and 5 minutes
- one group performance of an extract from a play
- one group performance of an original devised piece.

Good practice during your course

As you complete the practical work, get into the habit of discussing it and keeping notes on what you have done and why you did it in that particular way. Your teacher will be following your progress, so it will be important for you to show that you know and understand what you are doing. Most importantly:

- show commitment to your group and the project
- carry out research for yourself and share it
- listen to advice and criticism, and act upon it
- set targets to ensure the work progresses effectively.

The practical coursework performance is like any presentation in front of others. It is natural that you might feel nervous. Just think of nerves as being proof of your excitement and a sign that you want to do well. As a performer, take a few deep breaths and go for it – make it your best. When you try your hardest, nobody can expect more of you and you will even surprise yourself and gain confidence for next time.

Plan ahead

As soon as you start work on the pre-release material, it makes sense to start thinking about the kinds of question you will answer in an examination. From your very first lesson, you will have been learning new vocabulary and approaches to drama. Keep notes of these after each lesson. At times when you have completed a project, organise your notes in a way that makes sense and will fit in with the kinds of question you might be asked. You will be given ideas on what you will need to know in each unit and section. Whether you organise the points in lists, spider diagrams or even pictures will depend on what works best for you, but do keep up to date. A few notes made often will get you into good habits and it is actually easier than leaving it all to build up. These notes and plans will help you revise.

> **Useful tip**
>
> "You're so lucky to be able to perform like that," said a fan to a brilliant performer.
>
> "Thank you," replied the performer, "and the more I practise the luckier I seem to be."

> **Useful tip**
>
> Remember that drama is a practical subject and that you should study scripted plays from a practical point of view, using your skills and knowledge.

1.3 How this book works

Objectives

In this section you will learn:

- how to get the best out of this book
- about the units covered
- how the book tackles different skills.

How to get the best out of this book

Whenever you start a new course the amount you have to learn always seems daunting. Don't worry. This book is here to give you help and confidence.

- First, dip in and look at the units that interest you most.
- Notice that most units are arranged in pairs of pages called "spreads".
- Work your way through one spread at a time.
- Follow-up the suggestions and apply them to your own work.
- Take note of the various boxes on this spread, which are used throughout the book to help you and offer guidance.
- Key terms in orange are defined in a box on the page and/or in the Glossary at the end of the book.

Figure 1.2 Students preparing for a performance

Remember

All the examples given in this book can be applied elsewhere. Use the advice and skills information given and transfer it to your own work. Experiment and enjoy!

Activity

These boxes suggest activities you might do in the classroom or at home. Here is one to get you started.

1 Turn to Unit 2 and read the spreads on preparing your individual piece of work. You will discover how you might approach your own individual project. Turn to the costume design spreads in Unit 4. These will give you a start on considering costume design both for your coursework pieces and the written examination.

What this book covers

The book is arranged in sections to support you in all areas of your work.

Unit 2 Scripted work deals with approaches to acting and scripted work and gives advice on interpreting a text that you will find useful as actor or director.

Unit 3 Devising practical work covers a wide range of approaches using different stimuli such as you may find presented in the pre-release material. Many of the working methods, particularly improvisation, will also be helpful in preparing for scripted work.

Unit 4 Design and technical work explores a range of elements that can enhance a production and help communicate ideas effectively.

Unit 5 Developing your physical skills addresses movement, gesture, facial expression and voice – aspects that you will need to employ in your performances.

Unit 6 Preparing for a performance gives practical advice that will help you to present your performance more efficiently and effectively.

Unit 7 Writing about your performances offers advice to help you prepare for examination, and gives examples of the kind of questions you may be asked. You are strongly advised to look at this section early in your studies in order to make best use of your practical work.

Figure 1.3 A student performance

Useful tip
These boxes are written by experienced teachers. Each box will give you some helpful advice. The tip from this box is:
Always take notice of the blue boxes, they will point you in the right direction!

Remember
Drama is what you make of it. Though it may seem difficult at times, if you give it your best effort you will enjoy it and feel the satisfaction of having achieved something important. Good luck!

Key term
These boxes give definitions of words that may be new to you. You will also find these and other important terms in the Glossary at the end of this book. Learn them, as you will need to use them in your work. Learn to spell them correctly for written examinations.

Unit 2 Scripted work

Professional actors spend their lives becoming characters, or people who someone else has created. When you begin to work on a scripted play, you will be given many clues how to create and portray your role. Some of the characters you perform will have their roots in history and be based on the lives of real people. In this chapter we will be looking at how to create a character in a scripted play and the skills and techniques you can apply in order to create that role.

For Component 2, the coursework, you will produce:

- one individual piece (3 to 5 minutes long) comprising:
  - a performance of an extract from a play
- two group pieces (up to 15 minutes each), comprising:
  - one performance of an extract from a play
  - one original devised piece.

For this component you will use a play written by a playwright. You will learn the lines, then take the words and directions from the page through the rehearsal process to the stage, and finally perform it to an audience.

Objectives

In this unit you will learn to:

- consider how to create a role in performance
- explore how a playwright creates a scripted play
- study and perform a monologue and a scripted play
- consider period, society, culture and genre
- prepare for your written examination question.

2.1 Acting

What professional actors do in rehearsals is attempt to create the roles written by the playwright and the way they do this depends very much on the **period**, **society**, **culture** and **genre** of the play.

Period is the time when the play is set. It is very important to consider this, because the historical context of the play will have a direct influence on the way the characters behave.

Society deals with the people portrayed in the play, their identity, and how they lived in their community at this time. The type of characters and their relationships with each other also directly influence the way your character will behave on stage.

Culture deals with the ways of life, rites and rituals of the particular society represented in the play.

> **Key terms**
>
> **Period:** the time period in which a play is set.
> **Society:** who the characters in the play are.
> **Culture:** how the characters in the play live their lives.
> **Genre:** the type of production, for example, comedy, tragedy, thriller or documentary.

Figure 2.1 Actors from a production of The Crucible *by the Royal Shakespeare Company in London, UK*

Genre is the category that your play fits into. It could be a comedy, tragedy, thriller, documentary, melodrama or even sometimes a mixture of genres. Your group will decide the way in which your play is acted. This will be influenced by the nature of the play being studied and prepared for performance, and linked to the four Key terms discussed such as the historical period in which the play is set.

> **Useful tip**
>
> Your individual piece is called a monologue.

Getting started

When people talk about 'drama' they first think of acting plays, and to most people plays mean 'scripts'. When we go to see plays at the theatre, those plays usually begin life as a script. We, as an audience, experience the words and thoughts of the playwright as the actors breathe life into the characters they shape and create on stage before us.

You will perform an individual piece using an extract from a play. The role you create will be formed in discussion with your teacher and by studying the playwright's intentions for the character.

You will also perform a group piece using an extract from a play. The performance piece you create will be in collaboration with other members of your group.

Scripted work

2.2 Looking at a play

How do we begin to act out a scripted play? The first thing to do is to get some idea of the plot of the play and look at some of the characters the playwright has created for us. The cast list is at the beginning of the play, and this usually includes the cast of the first time the play was performed, as well as date and location. You will need to decide who will play which part in your group.

> ### Activities
> 1. Read through the cast list at the beginning of your chosen play. Imagine what the characters are like and see if you can get any clues from the playwright about the characters. Find out how old each character is. How has the playwright described the characters?
> 2. Look at the cast list of the first performance of the play you are going to perform. Where was it performed, and when?

Figure 2.2 Actors rehearsing from script

Once a play is cast, the process of rehearsals can begin. Rehearsals can take place over a period of time, but for the purposes of this piece of coursework, the rehearsal period will be approximately 10–12 weeks. Of course, you are not rehearsing as a professional actor does – you have other school commitments – but the process of putting on a play is the same. A great deal of time and hard work and a commitment to the other members of your group is essential.

Your rehearsal process will soon fall into a pattern which will probably include lunchtime and after-school rehearsals, but it will be a pattern that will suit your needs. You may not be required to rehearse at every meeting of the cast, particularly once the play has been **blocked**.

Key term

Blocking: being told by the director where to stand, move or sit as you move through the first reading of the play; you can make notes of these moves in your script to help you to remember them in the next rehearsal.

You, as a performer, must learn your lines and take direction from the person directing the play. In most cases, this will be your teacher. Now, you are a member of the cast of the play, and once you have been given the name of the character you are to portray, the fun can begin!

Figure 2.3 US director John Malkovich talking through his ideas during rehearsal

The **director** has the responsibility of pulling all the ideas of the cast members together to create the play. The Duke of Meiningen (1826–1914) is thought to be the first theatre director in Western theatre, and he developed many of the basic principles of stage drama used in modern theatre.

Modern theatre directors also have to work with advances in modern technology and special effects. Even in a school setting, these can be quite sophisticated and enhance the theatrical experience for the audience.

Remember
It is your individual acting skill that is being assessed, not the production as a whole.

Key term
Director: the person who tells an actor how and when to do something on stage.

Activities
1. Explore and research your character, finding out as much about them as you can. Did they actually live? In what time period, and where? This will help you when you come to portray your role.
2. The theatre directors Konstantin Stanislavski and Antonin Artaud were influenced by the principles developed by the Duke of Meiningen. Research these two directors and note your findings. How does their work influence your own performance?
3. Research a director of a non-Western theatrical tradition (e.g. Kabuki or Yoruba) and discuss your findings as a class. How might this tradition influence your performance?

2.3 Creating a character

Once you have cast your play, you will need to think about how to create your character. As an example, this chapter focuses on a play called *The Crucible*. **Please note that this play may be considered inappropriate in your school or cultural setting** as it focuses on the effects of religion and politics on a society, and contains strong themes. You may therefore wish to focus your scripted work study on another classic play, as the skills you learn and produce here are easily transferable to any scripted play. Even without in-depth study of *The Crucible*, you will still find the following pages helpful in guiding the process for creating a character.

The Crucible was written by American playwright Arthur Miller in 1953. It is set, however, at a much earlier time, in 1692, in a small community in Salem, Massachusetts. Their ancestors arrived from England to the New World on a ship named the *Mayflower*, but this community developed their own laws and rules for their society and their community to be established. They were 'Puritans' and had a strict code of conduct to live by.

Figure 2.4 The court scene in The Crucible *– note Mary Warren's costume*

When you begin to create your character, you must consider who you are in the play and what your relationship is with the other characters. *The Crucible* has a number of main characters. They include:

- Reverend Parris: mid-40s, the reverend of the town, ambitious and not well liked
- Abigail Williams: 17, strikingly beautiful, the niece of Reverend Parros
- Mary Warren: Abigail's friend – she is subservient, naïve and lonely
- John Proctor: mid-30s, attractive, strong, a pillar of the community
- Elizabeth Proctor: John's wife, a strict religious woman who has never lied
- Reverend Hale of Beverley: well-meaning but influenced, even blinded, by his beliefs
- Tituba: a slave and scapegoat.

Now we go about creating a character. For the purpose of this chapter, we will look at two characters: John Proctor and Abigail Williams.

Useful tips

The process of creating the character focuses here on two main characters, but the processes involved are the same for whichever role you are portraying.

You may be assessed on the creation of your character in your practical coursework. The playwright also gives lots of clues in how to develop your character. Look for the clues in the text you choose.

Creating the role of John Proctor

You will first have researched the background detail to *The Crucible* (or other chosen text) and you will have explored many aspects of the character as you have blocked the play. John Proctor's first entrance is in Act One. The doomed affair between Abigail and John is clearly established and the subject of much gossip within the community. His desire for contact with Abigail still persists, but it is forbidden as he is married. He has, in their view, committed a sin and is accused in open court later in the play.

Now we consider John Proctor's first entrance in the play. We have been given clues about him earlier in the scene when the girls were discussing the events in the forest, and how Tituba was creating a love potion for Abigail concerning John. Consider the language of the scene and the way the girls react to his entrance. They are afraid of him but also intrigued by him, particularly Abigail, who fully believes he has come to Salem to see her.

Next, follow the plan (see Figure 2.5) to help you to build up his character by exploring and creating a character profile. 'Key moments' are the significant events in the development of the character, that the director may choose to mark. In exploring his character in this way and analysing what attributes make up his characteristics, you are ready to begin the rehearsal process of developing the character through your research of John Proctor, the man.

Now look at the characteristics in Figure 2.6 to build up your character profile. As you begin to create his character, you will ask yourself the following questions:

- What does he do?
- How does he do this?
- Why does he do it in this way?

> ### Activity
> **Hot-seating** is a useful way to develop a believable character (see Section 3.8). Try a hot-seating exercise with John Proctor or another character from a play of your choice. The actor sits on a chair, in character, and class members can ask him/her questions about the character, either as themselves or also in role.

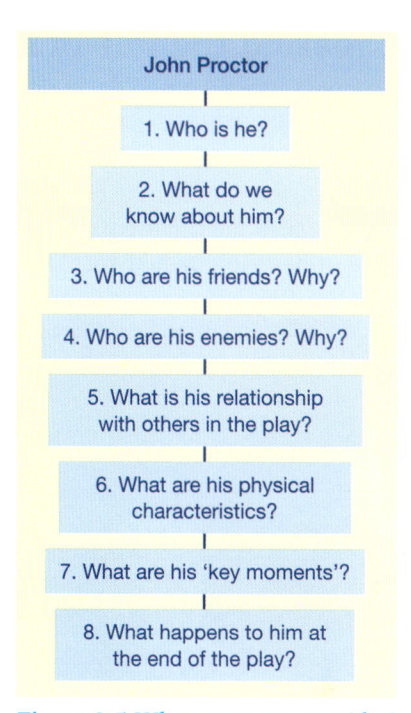

John Proctor
1. Who is he?
2. What do we know about him?
3. Who are his friends? Why?
4. Who are his enemies? Why?
5. What is his relationship with others in the play?
6. What are his physical characteristics?
7. What are his 'key moments'?
8. What happens to him at the end of the play?

Figure 2.5 What must we consider?

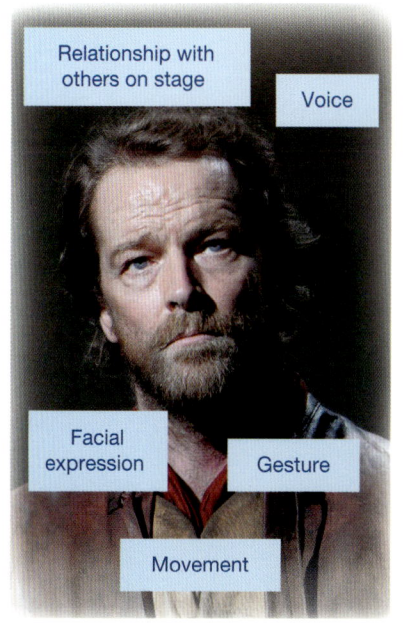

Relationship with others on stage · Voice · Facial expression · Gesture · Movement

Figure 2.6 Who is John Proctor?

Key term

Hot-seating: the technique of an actor staying in role while answering questions from the audience about the character's thoughts and feelings. The actor can involve the audience by asking them for advice.

Creating the role of Abigail Williams

The next character we are going to consider is Abigail Williams. Consider the language Abigail uses at the opening of the play. We see here a contrast between her conversation with Reverend Parris, and the conversation she has with her friends once he has left the room. She speaks with authority and manipulates the other girls with threats and ultimatums. The other girls are in awe of her. How would you show this when you are creating the role?

> **Remember**
> All the techniques explored for creating a character are transferable to whichever role you are playing.

Figure 2.7 Abigail Williams is a strong presence in The Crucible

Look at Figure 2.8 and Figure 2.9 to see how to create her character profile. This technique will also enable to you understand how a professional actor creates a role in performance.

In exploring Abigail's character in this way and analysing what makes her behave the way she does, you are now ready to begin the rehearsal process. Your detailed research will help you to create the role of Abigail Williams and you will know her before you begin.

In scripted drama there are no insignificant roles. The playwright has constructed the drama to accommodate a number of actors, and they all must play their part for the drama to unfold and the story to be understood by the audience.

All the other characters in *The Crucible* are important. They are responsible for supporting the story and for the development of the underlying themes of greed and revenge which help to shape the drama as it unfolds on stage.

In your performance, it is not necessary to have as many actors as there are roles. It is common for one actor to play more than one role. You will have to watch out for characters that interact with each other – but it can provide an opportunity for you to use acting skills such as voice and gait (see Section 2.4) to make the distinction between the two roles. Masks (see Section 4.4) may help, but it is your performance that will set the characters apart.

Activity

Choose a scene or section from *The Crucible* or a play of your choice. As a class or in pairs, discuss what your character is doing, how they are doing it and why they are doing it in that way.

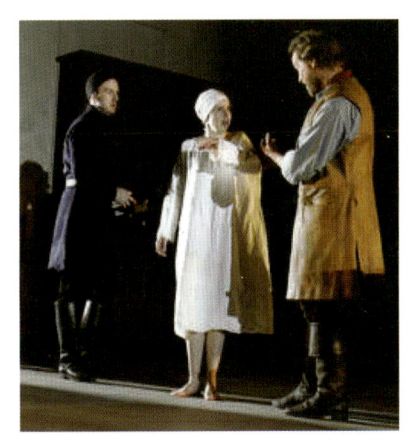

Figure 2.10 The characters John Proctor and Abigail Williams in The Crucible

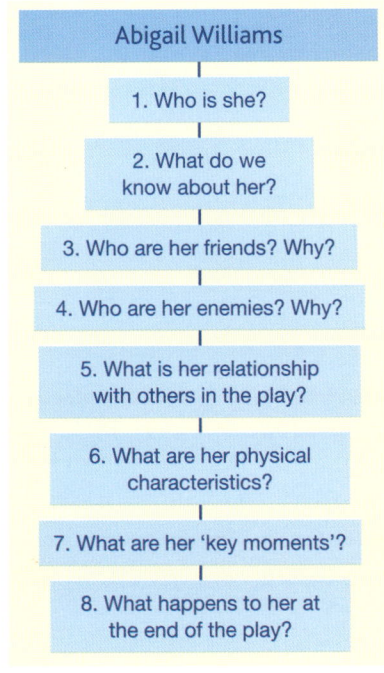

Abigail Williams
1. Who is she?
2. What do we know about her?
3. Who are her friends? Why?
4. Who are her enemies? Why?
5. What is her relationship with others in the play?
6. What are her physical characteristics?
7. What are her 'key moments'?
8. What happens to her at the end of the play?

Figure 2.8 What must we consider?

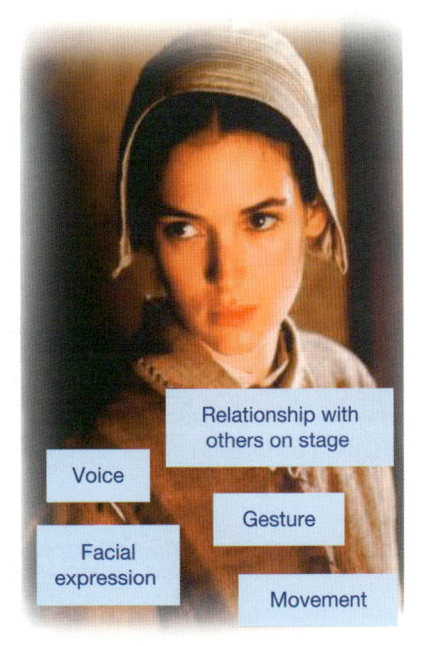

Figure 2.9 Who is Abigail Williams?

(Voice, Facial expression, Gesture, Movement, Relationship with others on stage)

Scripted work 15

2.4 Preparing your individual piece of work

Look ahead to Section 3.9 where there is detail of using a **monologue** to develop a character.

When you are working on a piece of scripted work, the character will have been created by the playwright and it will be your job to understand the **context**, the period, the **style** and genre and the character's relationships with other characters in the play.

You will need to be **fully aware of the whole play from which your extract is taken** so that you can give a truthful and believable performance.

In this speech from Shakespeare's *Twelfth Night*, Viola, alone on stage, has realised that Olivia has been taken in by her disguise, believes her to be a man and has fallen in love with her. You may find the language challenging – and for this reason, Shakespeare will not be included in the pre-release material for Cambridge IGCSE. However, try to follow Viola's changing thought processes throughout the extract.

> **Key terms**
>
> **Monologue:** when a character on stage speaks alone, sometimes directly to the audience.
>
> **Context:** the background information surrounding a play which helps us to understand the events and the characters.
>
> **Style:** the way in which the production is presented and performed (see Unit Section 3.16 for more on style).

VIOLA I left no ring with her. What means this lady?
Fortune forbid my outside have not charmed her.
She made good view of me... indeed, so much,
That sure methought her eyes had lost her tongue...
[For she did speak in starts distractedly.
She loves me, sure. The cunning of her passion
Invites me in this churlish messenger.
None of my lord's ring? Why, he sent her none.]
I am the man. If it be so, as 'tis,
Poor lady, she were better love a dream.
[Disguise, I see thou art a wickedness
Wherein the pregnant enemy does much.
How easy is it for the proper-false
In women's waxen hearts to set their forms!
Alas, our frailty is the cause, not we,
For such as we are made of, such we be.
How will this fadge? My master loves her dearly;
And I, poor monster, fond as much on him;
And she, mistaken, seems to dote on me.]
What will become of this? As I am man,
My state is desperate for my master's love;
As I am woman – now alas the day –
What thriftless sighs shall poor Olivia breathe?
O Time, thou must untangle this, not I.
It is too hard a knot for me to untie.

Twelfth Night by William Shakespeare from The Cut Shakespeare by Steve Gooch

In this version of the text from *The Cut Shakespeare* series by Steve Gooch, the sections in **bold** pick out the focus of the speech. This is a good starting point for working on an individual speech. It will define the through-narrative, or main story, and give a sense of the shape of the piece.

The first section of bold text shows the thought processes of Viola as she realises what has happened. In the second bold section, she states clearly the complicated situation she finds herself in.

In the final bold section, she explains the conflict and decides that she cannot solve the problem and must leave it to 'Time' to sort out; and this, of course, is what happens in the play that follows.

So how will you present the speech to the audience showing the conflict that is going on in Viola's soliloquy?

At first Viola is **thinking back** over the preceding scene.

- Where is your focus?
- What is your reaction?
- How are you feeling?
- Are you looking at the ring?
- Gazing out towards the audience?
- Half following the departing Malvolio to return the ring?
- Are you still in role as the boy Olivia thinks you are?
- What will be your stance and posture?
- Do you become more feminine as the speech goes on?
- When might this be and how will you show it?

Then realisation dawns and she **understands the feelings of Olivia** and how cunning she has been in arranging the meeting – 'She loves me, sure...'

- How will you show this?
- Is it spoken suddenly or thoughtfully?
- Are you shocked, amazed or amused?
- What will you do?
- Will you collapse onto the floor?
- Move downstage, closer to the audience, as though sharing a secret?
- Step back in shock?

Key terms

Conflict: an element of struggle, found in all drama; it may involve trying to resolve a problem or someone changing their life; it does not necessarily mean an argument.

Soliloquy: a dramatic device by which the inner thoughts and feelings of a character are revealed by an actor speaking aloud as if to him/herself.

Figure 2.11 Viola from a production of Shakespeare's Twelfth Night

She then speculates on the 'waxen' quality of women's hearts, which is ironic since she is a woman herself.

- How will you deliver this section?
- Wistfully? Bitterly? Forthrightly?
- Are you sitting, or wandering from stage right to stage left?
- What about gesture?
- Are your hands on your head or on your hips? Or where? Why?

Viola then **directly states the problem**. Olivia loves her, as a boy; and she, as a woman, loves Orsino.

- How is this delivered?
- Out front to the audience?
- To the heavens with eyes upraised?
- Quietly looking down at the ring?

In what mood should Viola **exit at the end of her soliloquy**?

- Are you optimistic or not?
- Do you stride off or sidle off?
- Is your tone of voice lively or resigned?
- Where are you going?

The importance of pace

Notice that the pace of Viola's monologue distinctly changes throughout the speech. If you were running in a long-distance race, you would not want to keep at the same speed or pace throughout – too slowly and you will come in last; too quickly and you will soon be tired.

> **Useful tip**
> Mark the moves you are making in your script. Use a pencil so you can change your mind as you rehearse.

> **Remember**
> It is easier to learn lines if you tie them into the action.

Figure 2.12 Viola and Olivia from Twelfth Night

It is important to remember this when planning your drama work. The speed of the action needs to be varied to fit the mood of the moment and the overall structure. Fast, exciting and energetic dialogue and action are hard to sustain and can be tiring to watch. Think carefully about creating changes of pace. A powerful and dramatic end may be given greater impact if it comes after a moment of quiet calm, while a loud and busy opening scene can grab an audience's attention, so that you can then change to a slower pace with quieter voice in order to share important plot details.

Get into the habit of discussing pace with others in your group and remember to keep notes for your reference. This will help you to give shape to your material and improve its impact on the audience.

Using your space

The extract below is a modern adaptation of the Greek tragedy *Antigone*, written by French playwright Jean Anouilh. In this version, the character of the Chorus is played by one actor. He is an all-seeing commentator on the action in the play (as is the chorus in a Greek tragedy). The Chorus, in this long speech to the audience, uses **direct address** to describe who everyone is and what will happen in the play. Note here the stage directions, which explain where each actor should be on the set.

Antigone, *her hands clasped round her knees, sits on the top step. The* three guards *sit on the steps, in a small group, playing cards. The* Chorus *stands on the top step.* Eurydice *sits on the top step, just left of centre, knitting. The* Nurse *sits on the second step, left of* Eurydice. Ismene *stands in front of arch, left, facing* Haemon, *who stands left of her.* Creon *sits in the chair at right end of the table, his arm over the shoulder of his* page, *who sits on the stool beside his chair. The* messenger *is leaning against the downstage portal of the right arch. The curtain rises slowly; then the* Chorus *turns and moves downstage.*

> CHORUS: Well, here we are.
>
> These people are about to act out for you the story of Antigone.
>
> That thin little creature sitting by herself, staring straight ahead, seeing nothing, is Antigone. She is thinking. She is thinking that the instant I finish telling you who's who and what's what in this play, she will burst forth as the tense, sallow, wilful girl whose family would never take her seriously and who is about to rise up alone against Creon, her uncle, the King.
>
> Another thing that she is thinking is this: she is going to die. Antigone is young. She would much rather live than die. But there is no help for it. When your name is Antigone, there is only one part you can play; and she will have to play hers through to the end.
>
> From the moment the curtain went up, she began to feel that inhuman forces were whirling her out of this world, snatching her away from her sister, Ismene, whom you see smiling and chatting with that young man; from all of us who sit or stand here, looking at her, not in the least upset ourselves—for we are not doomed to die tonight.
>
> CHORUS *turns and indicates* HAEMON.
>
> The young man talking to Ismene—to the gay and beautiful Ismene—is Haemon. He is the King's son, Creon's son. Antigone and he are engaged to be married. You wouldn't have thought she was his type. He likes dancing, sports, competition; he likes women, too. Now look at Ismene again. She is

> **Key term**
>
> **Direct address:** an actor speaks directly to the audience.

certainly more beautiful than Antigone. She is the girl you'd think he'd go for. Well… There was a ball one night. Ismene wore a new evening frock. She was radiant. Haemon danced every dance with her. And yet, that same night, before the dance was over, suddenly he went in search of Antigone, found her sitting alone—like that, with her arms clasped round her knees—and asked her to marry him. We still don't know how it happened. It didn't seem to surprise Antigone in the least. She looked up at him out of those solemn eyes of hers, smiled sort of sadly and said "yes." That was all. The band struck up another dance. Ismene, surrounded by a group of young men, laughed out loud. And… well, here is Haemon expecting to marry Antigone. He won't, of course. He didn't know, when he asked her, that the earth wasn't meant to hold a husband of Antigone, and that this princely distinction was to earn him no more than the right to die sooner than he might otherwise have done.

CHORUS *turns towards* CREON.

That grey-haired, powerfully built man sitting lost in thought, with his little page at his side, is Creon, the King. His face is lined. He is tired. He practises the difficult art of a leader of men. When he was younger, when Oedipus was King and Creon was no more than the King's brother-in-law, he was different. He loved music, bought rare manuscripts, was a kind of art patron. He would while away whole afternoons in the antique shops of this city of Thebes. But Oedipus died. Oedipus' sons died. Creon had to roll up his sleeves and take over the kingdom. Now and then, when he goes to bed weary with the day's work, he wonders whether this business of being a leader of men is worth the trouble. But when he wakes up, the problems are there to be solved; and like a conscientious workman, he does his job.

From the opening scene of *Antigone,* by Jean Anouilh

Figure 2.13 Antigone from a production of Jean Anouilh's adaptation of Antigone

The style of the writing in this speech is conversational and is directed at the audience. The Chorus makes no pretence that this is real life and the outcome is clear. He is not sentimental about what is going to happen but he needs to engage the audience from the very beginning.

Where will you be on stage?

The playwright suggests that the Chorus starts upstage with his back to the audience and then turns and moves downstage, to be close to the audience and to be able to speak directly to them. This would help him to build his relationship with them, as he will be their guide throughout the action.

You may decide to stay upstage so that you can be close to Antigone as you introduce her, or you may start downstage and circle back upstage towards her. It may be repetitive to move towards each character as you speak about them; note that the playwright leaves the Chorus in the same position throughout this speech, indicating the characters and focusing on the way he speaks to the audience. You may prefer to vary the positions on stage to keep the attention of the audience.

How will you move?

The Chorus has an easy-going manner but there are moments when he is conjuring up moments of action such as the meeting of Ismene and Haemon at the ball. Here you may need to think about a livelier **gait**, which you might contrast with a more sober approach as he describes the life and history of Creon, the King.

Perhaps he will sit by him or stand behind him. Remember that the actors are in **tableau** and the Chorus may move amongst them.

How will you speak?

The Chorus begins by outlining Antigone's appearance and thoughts. What will his tone be? Look at the language: there is tension and an urgency about what she is about to do. His voice is contrasting with her stillness.

There is a change in pace as he begins to introduce Haemon. He is almost gossipy here and maybe suggests that there is some friction between Ismene and Antigone. His tone may change, deepen and slow when we hear how Haemon suddenly asked Antigone to marry him.

When he tells the audience about Creon, his voice could reflect the age and exhaustion of the old man, then becoming lighter when his youth is described, and going back to a weary tone when outlining what has to be done.

> **Key terms**
>
> **Gait:** the manner in which a character walks and moves around the stage.
>
> **Tableau:** the character/s, clearly showing their role, appear as a still image.

> **Remember**
>
> Whether you are working on your speech as an actor or as a director, you need to be very clear about the journey you are taking the audience on and how you are going to grab and hold their interest throughout.

Activities

1 Select the opening scene from a play of your choice and draw a diagram showing where the characters will be on the set.

2 As a class or in pairs, choose one character from the scene above and discuss how he/she will move and speak.

Unit 3 Devising practical work

3.1 Devised work – an introduction

In addition to using well-known playscripts, actors are often involved in creating original pieces of work, known as devised theatre. As a group, you will be assessed on a performance of an original devised piece based on a **stimulus** of your teacher's choosing. There are many different stimuli which can inspire your piece of original work, and which will give your group a wide scope to develop and practise the group's performance and design/technical skills.

> **Objectives**
>
> In this unit you will learn to:
>
> - create a performance through a devised thematic approach
> - perform work before an audience
> - communicate your concepts and ideas to the audience
> - show awareness of different viewpoints surrounding a theme.

> **Key term**
>
> **Stimulus (plural stimuli):** the starting-point for a devised work; the idea, image or object that sparks off your work.

Getting started

There is no better way to develop an understanding of professional theatre companies' devised work than by doing some research. If you look at the brochures describing the work of professional theatre companies, you will find many examples of companies which devise and perform their own work. Devised work often includes:

- an **ensemble** of performers with a variety of talents
- a theme (for example, life and health)
- text
- musical elements
- movement elements.

All these are *characteristics of devised work* and will result in brand new and exciting work. Companies that produce this vibrant and thoughtful work all devise in a different way and use different starting points. They take account of the *skills* of the company, and so will you when you come to devise your own work. They also have a real passion to investigate and develop their stimulus. The starting point or theme is chosen because it is something that the group feels strongly about, and they have a viewpoint that they wish to communicate to an audience. When you choose your starting point, make sure that you are committed to the points you want to put across to the audience.

Here we provide details of a few with links to their websites, but you may choose to look up some theatre companies closer to you:

- Tamasha Theatre produces new plays inspired by the diversity of the global world. *The Arrival* is based on an illustrated novel by Sean Tan and incorporates exciting circus skills into the production. Another piece called *Lyrical MC* was produced by Sita Brahmachari using **verbatim** text from different schools. Look at their website to read more about their past productions.
- New International Encounter (NIE) is an international company which produces work using a variety of languages, musicians on stage and many theatrical styles. Their work *The End of Everything Ever* uses true stories and accounts of 'Kindertransport', the rescue of Jewish children from the Nazis in the Second World War, and follows the trials of one child. Look at their website to investigate their work.

Work based on well-known plays is also devised by theatre companies:

- Frantic Assembly's adaptation of *Othello* mixes movement, design, text and cinematic soundtrack to investigate the themes of prejudice, danger and fear. You can find out about how this company develop their work by looking at their website and their resource packs, especially the one for *Stockholm*.
- The Nuffield Theatre Company have devised a new piece *The Coast of Mayo*, inspired by their own production of *The Playboy of the Western World*. The actors had the experience of working on both the scripted and the devised piece.

Key terms

Ensemble: a group of people working together; everyone makes an equal contribution and there is no 'starring role'.

Verbatim theatre: the words of a real person are recorded as part of the research for a piece of work. The words used – and every pause or "er" – are then spoken by the actor in exactly the same way in the performed piece.

Did you know?

Devised theatre is sometimes called 'collaborative creation', particularly in the US.

Links

www.nie-theatre.com
www.franticassembly.co.uk
www.nuffieldtheatre.co.uk
www.tamasha.org.uk/productions/past
www.sharedexperience.org.uk

Activity

Research a devised production from a theatre of your choice, and note down your findings. What do you think is the main stimulus for the production? What are the characteristics of this piece of work?

Devising practical work

3.2 Resources and skills

Resources

Before you start thinking about your devised work, you will need to think about the resources and skills available to you which will influence the work that you produce. As well as all the printed material you have access to, your greatest inspiration will be the live theatre you have seen.

When you read a novel, it has an effect on your own writing; music you hear and art you see influence your work in these areas too. In the same way, productions you see inspire your own drama, especially in devised work where you can use a variety of styles and ideas. We have provided a few ideas here, but in every production there will be something which works really well, and very often an idea you can use and develop when you get back into your own performance space.

Small-scale touring productions are designed to fit into a small performance space and pack into a van, so they may provide simple ideas of how to create different scenes in the space and with the resources available to you in your own setting. A *composite set* can be used in many ways and lit to suggest different locations.

A *gauze* is another simple tool that you can use effectively in your setting, and which is not too expensive to buy. Change a scene downstage for one upstage of the gauze by *cross-fading* lights from downstage to upstage. Try projecting onto it. Notice how projections or video or PowerPoint slide show inserts are used in productions.

Experiment with entrances and exits and try out your work in a variety of stage forms (see Section 4.7).

> **Useful tip**
>
> When you get into your performance space, try to reproduce moments from a production that worked well. This will help you to remember what you saw when writing your notes and to realise that you can reproduce ideas you have seen.

> **Key terms**
>
> **Composite set:** a set on which it is possible to perform scenes in different locations or rooms in the same building without having to change the set all the time.
>
> **Gauze:** (scrim) a coarse-weave fabric which appears transparent when the scene behind it is lit; sharkstooth is the most opaque.
>
> **Cross-fade:** when one lighting state fades down as another one fades up, either instantly or over a period of time.

Figure 3.1 Still from Some Trace of Her *by director Katie Mitchell, National Theatre, London, UK*

Skills

Your other main resource is the skills of the people in your group, so start off by writing a CAN DO list and add everything you have learnt in all the drama you have done. Don't leave anything out, as it could all be useful in devised work.

Here is an example of a CAN DO list:

> - Effective tableau – could re-create a photograph or comic strip effect
> - Physical theatre – could be used in a comedy sequence
> - Stepping out from the action – direct address to the audience to make a point or move the action on
> - Improvising naturalistic scenes – to show the development of a character
> - Performance of text
> - Acting and interpreting a character
> - Dance or movement sequences – to create a mood or tell a story
> - Mime
> - Use of masks
> - Theatre from other periods – Greek chorus work, Commedia dell' arte
> - Different narrative techniques – narrator, flashbacks, interviews.

Key term

Devising: planning a production and working out how it can be performed effectively.

You will have many more to add to your own list, but do not forget individual skills that members of your group may have, and see if you can use them in your work.

There may be singers and musicians amongst you, or dancers and jugglers. Someone may have photographic skills which could be used to create projections, or someone may be able to make and edit video effectively. There are many ways you could incorporate these skills, for example:

- a solo violin could evoke a poignant moment
- tap-dancing a rhythm could suggest gunfire
- a stand-up comedian would be great at directly addressing the audience
- martial arts experts could choreograph a movement sequence.

There is no end to the skills you could use in **devising** your new work.

Figure 3.2 Devised work by *New International Theatre*

Devising practical work 25

3.3 Devising work from a theme

There are many themes that can inspire your devised work, so you need to make sure that all your group is interested in the theme you choose, and that you have something to say on the subject. It is important that you put your ideas across to the audience, using original work.

Perhaps you will be inspired by a poem you have enjoyed or by a group of poems on a theme that you are studying. Maybe a novel you are reading suggests a theme that you feel could be developed theatrically, or an issue dealing with environmental matters moves or concerns you. An article in a newspaper or a series of letters on a problem page in a magazine could all be the basis for a unique piece of theatre. In all aspects of your life be aware of ideas which you could investigate.

In the activities, you will see examples of popular plays and themes, and some suggestions to get you started.

Activities

1. Investigate the theme of **childhood** and changing relationships of friends and enemies.
   - Try a movement sequence based on playground games to get you started.
   - Read the poem *Tich Miller* by Wendy Cope, which investigates the theme of fitting in at school.

2. Investigate the theme of **war**.
   - Look at *Blue Remembered Hills* by Dennis Potter. Use sections of this play to inspire a piece on war and its effects on those at home.
   - Improvise some scenes as the parents of the children in the play.
   - Look at *Spies* by Michael Frayn for another view of children during the war. This is written as a **flashback** from the point of view of a character now grown up.
   - Try to flash forward from the end of *Blue Remembered Hills*. Improvise scenes of the children in the future facing up (or not) to what they have done.

3. Look at *The Caucasian Chalk Circle* by German playwright Bertolt Brecht. Investigate the theme of **war** from the point of view of the refugee. Grusha has to leave her home because of the war and has to keep a child safe.
   - Research this situation in the modern world – you could include video material of war situations and audio reports of what people think about refugees. You may find material in newspapers or someone you know may have personal experiences to relate.
   - Compare this to Grusha's situation – begging for milk, escaping from soldiers, being rejected by her family.

Key term

Flashback: when the narrative in a drama switches from the current time (current from the point of view of the characters) to an incident from the past, perhaps as a memory or a dream by one of the characters.

4. Another theme you could work on is that of **adoption**. *The Caucasian Chalk Circle* asks the question "Who is the real mother?" and looks at the position of an outsider.

   Find another play that deals with the theme of adoption, and compare how the theme is explored in both pieces. Consider the following questions:

   - How does the character feel about being adopted?
   - Is the play portraying adoption in a positive or negative way?
   - What other themes does each play present?

Useful tip

Keep a research file on everything that could be included in your work. While it's better to have too much material than too little, do not be afraid to discard some of it.

Figure 3.3 A Marathi performance of The Caucasian Chalk Circle

In *Sepia and Song* by David Foxton, there are two plays which use the documentary style to put across fact and opinions. *I was a Good Little Girl 'til...* is about the suffragettes struggling for the vote in the UK, and *Titanic* investigates the sinking of the ship.

Oh! What a Lovely War (Theatre Workshop) is a wonderful example of documentary-style theatre about how the First World War resulted in the deaths of millions. If you are attempting documentary work, it is a good idea to look at this play to see how the newspanel is used together with projections, songs from the period, and acting and dancing to make its points.

Key terms

Documentary: putting factual information across to the audience.

Newspanel: written messages flashing across a screen during the play, perhaps giving facts that the audience would find difficult to take in if they couldn't see them.

Devising practical work

Devising work on a theme: 'Despair'

To give you a feel for devised work, take a basic theme and split up the class to produce a short piece involving everyone and using a variety of skills. One example of a theme is the single word 'despair', which gives plenty of scope for creativity.

You may need to divide up into small groups. One group could work on each of the following:

- a simple set which suggests the theme to the audience
- lighting which will evoke the mood, using equipment such as a **gobo** or a **gel**
- a soundtrack which can include vocal, live, or recorded sound
- images to be projected as a slide show, using equipment such as a gobo or gel – an overhead projector can be useful if you don't have a computer-driven projector
- a costume which will enhance the theme and which someone in the group will wear – use anything you can easily get your hands on
- collecting or making suitable props
- design make-up or make a mask
- a movement/dance sequence
- some text
- a rehearsed, improvised scene.

You may also add anything else you might like to include. This will depend on the interests of your class but try to cover as many as possible and have at least two people in each group.

Some points to consider

If you are working on the set, you will have to define your performance space and use whatever is available to you.

- Look especially at fabrics and think about the effects of colour and texture.
- Experiment with creating a variety of levels, using a **rostrum** for example, to give useful space for the performers.

The lighting team will need to consult with the set designers.

- Try to use the minimum number of lanterns, light from unusual angles.
- Try out gobos and different coloured gels.
- You might try using only torches.

It might be useful to create your costume on a member of your team and then you will know immediately what it looks like.

- Make sure it is possible to get on and off and is comfortable to wear.
- Experiment with a variety of materials, think about spraying and painting, and do not forget accessories.

> **Key terms**
>
> **Gel:** a coloured film placed in front of a lantern to change the colour of the light on stage.
>
> **Gobo:** a small, perforated metal sheet placed between the lamp and the lens of a spotlight to project an image on to the stage.
>
> **Rostrum (plural rostra):** a portable platform which you can use to create interesting levels.

> **Useful tip**
>
> It might be a good idea to try working in a skills area you haven't tried before. This will help to widen your skill sets, and give you a better understanding of the skills required for different elements of a production.

Some examples of text

You can use these examples to inspire your work on the theme of despair.

Chorus speech from *King Oedipus*

Sorrows beyond all telling:

Sickness rife in our ranks, outstripping

Invention of remedy – blight

On barren earth,

And barren agonies of birth –

Life after life from the wild-fire winging

Swiftly into the night.

Sophocles, trans EF Watling

An extract from *Metamorphosis*

What an exhausting job I've picked on! Travelling about day in, day out. It's much more irritating work than doing the actual business in the warehouse, and on top of that there's the trouble of constant travelling, of worrying about train connections, the bed and irregular meals, casual acquaintances that are always new and never become intimate friends.

Franz Kafka, trans Martin Secker & Warburg, Schocken Books Ltd

Structuring your work

You will need to take time to look at the contributions of all members of the class and then make decisions about how it will be put together. Your teacher will probably be the best person to take a directorial role at this point and shape the work.

The possibilities are endless:

- The movement team could teach their sequence to other people and their work could be an opening or closing sequence.
- You could start in blackout and fade up onto a costumed figure or an empty set.
- Is the soundtrack playing at this point? Is it playing throughout the whole piece or at significant moments?
- Is a dancer wearing a mask or is an actor appearing in appropriate make-up?
- How will the text be delivered – by actors on stage, by a single voice or as a chorus? Will the presentation be static or accompanied by movement?

When your work is **structured** you will need to rehearse this carefully as it may be quite a technical production. You will need to do a full technical and dress rehearsal before you present your work to an audience.

It is a good idea for everyone to keep a full record of the order of the work and the cues. As you will all be involved in the performance, get someone to record it so that you can look at it later.

> **Key term**
>
> **Stucture:** the organisation of work in terms of its starting point, its setting and narrative line.

> **Useful tip**
>
> Keep a full record of the process as it will be helpful when you create your own group work.

3.4 Devising work using a poem

A poem may present a theme for you to investigate, which you will find challenging and which will give you a chance to develop and improve your skills.

Try the theme of money inspired by the poem *Money (rant)*, by Benjamin Zephaniah. Here is the first verse, which has many ideas. If you wish to read further, the whole poem is packed with themes which you could develop.

Money (rant)

Money mek a Rich man feel like a Big man
It mek a Poor man feel like a Hooligan
A One Parent family feels like some ruffians
An dose who hav it don't seem to care a damn,
Money meks yu friend become yu enemy
Yu start see tings very superficially
Yu life is lived very artificially
Unlike dose who live in Poverty.
Money inflates yu ego
But money brings yu down
Money causes problems anywea money is found,
Food is what we need, food is necessary,
Mek me grow my food
An dem can eat dem money.

First stanza from 'Money (rant)' by Benjamin Zephaniah

Figure 3.4 *The poet Benjamin Zephaniah*

> **Useful tip**
>
> Collect any images and material that will help. Everyone should bring at least one stimulus, for example, a photo, magazine/newspaper article, poem or piece of music.

Now that you are working in a smaller group, you will need to be able to use a variety of *performance skills* and work in different areas. If you have a particular skill, try to find an opportunity to use it.

> **Useful tip**
>
> Look at your CAN DO list and experiment practically with your ideas. Choose a suitable performance style for each idea.

Activities

1. a Consider the rhythm of the poem – will you use this piece in your work? Try to create a movement or dance piece using the rhythm, or write a song on the theme.
   b Look at the line: 'Money meks yu friend become yu enemy.' There is scope here for an interpretation through dance, improvised scenes, monologues showing changes in relationships, or poems based on the theme.
   c The last three lines could lead you into environmental issues.
2. a Look at other related scripts, such as the 12 August scene from *Oh! What a Lovely War*, in which the characters discuss making money from war; or the envy caused by land ownership in *The Crucible*.
   b Use scenes from *The Good Person of Szechuan* by Bertolt Brecht. Can you be 'good' if you have money?
   c Make a large spider diagram to record all your ideas at this stage.

Structuring your work

When you have tried out all your ideas, you will have to organise your separate sections into a piece of work which will keep your audience interested. You will also need to make sure that the point you want to put across is clear. What is your work about? What is your concept?

Throughout the preparation period, you need to keep careful notes and records of what you have done. You could keep a written notebook or a series of photographs. You may choose to use sticky notes or looseleaf sheets which you can swap around until you've got the order right. A slide show of scenes may help to sort out the order of the scenes, or even pieces of paper pegged onto a washing line.

Aim for variety in presenting your ideas but make sure that the **transitions** between your scenes or sections are smooth and do not hold the action up. You can do this by having a composite set (see Section 3.2), planning where any costume changes will happen, and making sure that something else is happening on stage while changes are taking place. Use **cross-fades** for your lighting rather than blackouts. Save blackouts for when you need them for dramatic reasons.

Make sure that your opening really engages the interest of the audience. Shakespeare did: for example, look at the fight scene in Romeo and Juliet. It doesn't have to be a very busy scene: a solo singer, alone on the stage, or a monologue spoken directly to the audience can be just as engaging.

At the conclusion of your work, make sure the point you are hoping to make is clear. What are you hoping to say about money and its effects on people?

- Does it make people selfish?
- Does it cause rifts between family and friends?
- Does it make people create inferior work, just for the money?

It's your work and it is unique and meaningful to you and your group, and it is now your work to make a unique and meaningful experience for your audience!

Figure 3.5 A solo actor on stage

> **Remember**
> Everyone in the group should take responsibility for keeping a record of the process and have a copy. You may want to use this work to answer Section A on the written paper.

> **Key terms**
> **Transition:** a change between scenes or sections of your work.
> **Cross-fade:** when one lighting state fades down as another one fades up, either instantly or over a period of time.

Poetry and voice work – *Night Mail*

This section aims to help you feel confident to use a poem as a stimulus for your devised coursework and to help you to develop your vocal, interpretation and performance skills.

The stimulus for this section is W. H. Auden's poem *Night Mail*. Written in 1936, the poem was part of a documentary on the work of the mobile sorting offices that took the overnight mail from London, England, to Glasgow in Scotland. Your teacher may have access to a copy of the film or you may be able to view it online as part of your research. The poem was set to music by the British composer Benjamin Britten who was, at that time, a newly emerging talent. You may wish to research W. H. Auden and Benjamin Britten as part of your background work.

Figure 3.6 *The poet W. H. Auden (on the right) in the 1930s*

> ## Activities
> 1. Warm up: Before you start you will need to do a voice warm-up. Choose some exercises from Section 5.3 'An introduction to voice work' to get you started. The "Hoo, hoe, hee" vowel exercise will have you breathing deeply and feeling energised. As the poem is about a train journey, you may wish to use consonants that sound like train wheels on rails. "Du-du-du-da" and "Tu-tu-tu-ta" are good starting points.
> 2. Reading the poem: Once you have warmed up, read the poem carefully. You will see that it is presented in four sections. Look at each section and note down what makes it different from the others. The differences may be to do with images, rhythms or mood.

Night Mail

I

This is the Night Mail crossing the border,
Bringing the cheque and the postal order,
Letters for the rich, letters for the poor,
The shop at the corner and the girl next door.
Pulling up Beattock, a steady climb:
The gradient's against her, but she's on time.
Past cotton-grass and moorland boulder
Shovelling white steam over her shoulder,
Snorting noisily as she passes
Silent miles of wind-bent grasses.

Birds turn their heads as she approaches,
Stare from the bushes at her blank-faced coaches.
Sheepdogs cannot turn her course;
They slumber on with paws across.
In the farm she passes no one wakes,
But a jug in a bedroom gently shakes.

Section I task

The first section of the poem conjures images of a train steadily running through open countryside. Notice the pairs of rhyming couplets and the underlying rhythm that makes this section strong and muscular. Read it aloud and then see if you can get the effect of a powerful train by having a background support of vocal effects from others in your group. Experiment with voice exercises using 'D' and 'T' sounds as described in the Warm Up activities in Unit 5. Perform the section to the other groups in your class.

> II
>
> Dawn freshens, the climb is done.
> Down towards Glasgow she descends
> Towards the steam tugs yelping down the glade of cranes,
> Towards the fields of apparatus, the furnaces
> Set on the dark plain like gigantic chessmen.
> All Scotland waits for her:
> In dark glens, beside pale-green sea lochs
> Men long for news.

Did you know?
When *Night Mail* was written, the rails were not welded together end-to-end as they are now, so the wheels made a rhythmic sound as they crossed the gaps.

Figure 3.7 The Royal Mail train as shown in the 1936 documentary, Night Mail

Section II task

Read the poem aloud, perhaps taking a line each as you go round your group. In what ways does Section II differ from Section I? The next exercise may help you.

In pairs, sit back-to-back with a copy of the poem each. Person A will be the reader and will have to think about how to perform it. Person B's job is to pick out all the visual-image words in this section of the poem; he or she will need to underline these words while Person A is preparing.

When you are ready to start, Person A will read Section II of the poem while Person B will say the underlined words at the point at which they arise. The effect will be of certain words being *stressed* as both people say them.

Devising practical work

This works on the basis that two heads are better than one and Person B is helping Person A to be aware of the images in the poem. You can use this technique for any piece of text you have to work on.

If you want to fit background vocal sounds to this section of the poem, think about what kind of sounds are the most appropriate. Is it as mechanical and rhythmical as Section I?

Does it have a different mood? What mood is it? How will you convey this to your audience?

> **III**
>
> Letters of thanks, letters from banks,
> Letters of joy from the girl and the boy,
> Receipted bills and invitations
> To inspect new stock or to visit relations,
> And applications for situations
> And timid lovers' declarations
> And gossip, gossip from all the nations,
> News circumstantial, news financial,
> Letters with holiday snaps to enlarge in,
> Letters with faces scrawled in the margin,
> Letters from uncles, cousins, and aunts,
> Letters to Scotland from the South of France,
> Letters of condolence to Highlands and Lowlands
> Notes from overseas to the Hebrides,
> Written on paper of every hue,
> The pink, the violet, the white and the blue,
> The chatty, the catty, the boring, adoring,
> The cold and official and the heart's outpouring,
> Clever, stupid, short and long,
> The typed and the printed and the spelt all wrong.

Figure 3.8 A postal worker from the 1930s

Section III task

This section of the poem may be described as being like an early rap because of the rhythms and the way the words are crammed in together. Each word needs to be spoken clearly and crisply, and each consonant must be sounded. Notice the way in which the verse forces you to change the pace of your delivery. Make notes on where it speeds up and where it slows down. Which section gives the impression that the train is moving at its fastest?

The next exercise is a lot of fun and should help you with both expression and vocal control. Walk round the room with a copy of the poem, reading Section III aloud and being careful not to bump into anyone. When you come to a word you think is important, make a big gesture as you say that word. For example, if you choose 'faces' you may decide to pull a silly face as you say it. This is not an exercise that translates directly into your performance; instead, it is a way of having fun with the words. You should find that the next time you read the poem from the page, it will have more life as you have created an imaginative connection to the words you are saying. Try it and see what effect it has on your vocal performance next time you read the poem aloud.

> **Did you know?**
>
> The Hebrides, which W. H. Auden mentions in *Night Mail*, is an archipelago, or group of islands, off the west coast of Scotland.

IV

Thousands are still asleep
Dreaming of terrifying monsters,
Or of friendly tea beside the band at Cranston's or Crawford's:
Asleep in working Glasgow, asleep in well-set Edinburgh,
Asleep in granite Aberdeen,
They continue their dreams,
But shall wake soon and long for letters,
And none will hear the postman's knock

Without a quickening of the heart,
For who can bear to feel himself forgotten?

By W. H. Auden

Figure 3.9 A Scottish city at dawn

Section IV task

In this section of the poem, the train journey is winding down. Think about the way in which the pace of your delivery will have to reflect this. You are given images of the people who will soon be receiving the letters that the night mail is bringing. Look back at the exercises given for Section II as these will also work well here. Think about the main images and the most appropriate tone of voice. The night mail and its crew have done their job, but what about the people who will benefit? Do they even know the effort that will have gone into delivering their letters? What impact does the last line have on you?

Group performance

When you have completed the tasks for each section, think about how you can put them all together into a performance. This may involve your group taking responsibility for a single section or the whole poem. Think about what you want the piece to communicate to your audience. Think about the style of presentation you will have to adopt in order to convey your **directorial concept** most effectively.

The images that come into your head when you read the poem can start a train of thoughts and stories. Pool ideas and bring something of yourself to the piece so that you create something new that belongs to you and your group. Each group can present their work to the others and performances can be compared. Alternatively, you can combine your efforts to produce a performance involving everyone in your class. Experiment and try different approaches to find out what works best. You may wish to use parts of the poem as a background to a devised piece entitled *Waiting for a Letter,* or present the poem using just vocal or percussion sounds alongside the words. The techniques and skills you have learnt here can be applied to your next performance project.

> **Key term**
>
> **Directorial concept:** the basic ideas that underpin your production and that you want to communicate to the audience.

> **Useful tips**
>
> You can develop your work further by looking at other poems with railway settings, for example:
>
> - *Addlestrop* by Edward Thomas
> - *The Whitsun Weddings* by Philip Larkin.
>
> You may also like to read *The Travelling Post Office*, an Australian ballad by A. B. 'Banjo' Paterson, the author of *Waltzing Matilda*.

Devising practical work

3.5 Performance skills

When you perform your work you will want to give your best performance. It's a good idea therefore for your director to provide advice throughout the preparation period. The aim is to communicate your concept to an audience, so the comments of your fellow group members or classmates – as an audience themselves – will be very helpful. The advice from your teacher or director will be invaluable.

You can get tips and help from your director about:

Characterisation

- Is your character believable?
- Is your character's relationship to other characters on stage believable?
- Are you sustaining your characterisation throughout?

Use of voice

- Are you speaking clearly and at the right volume?
- Is the language you are using appropriate?
- Is your accent accurate and fitting to the role?

Use of movement

- How are you showing the mood of your character?
- How are you showing the status of your character?
- Are you in the right position on stage?

Figure 3.10 An actor uses his voice and body to convey a character

You may find that you are needed to play a variety of roles in this work. You will need to concentrate throughout and be completely prepared. You will have to be in character from the moment you enter the scene. Have a clear written outline of the order of scenes during the rehearsals, and be in the right place at the right time.

If you are the director, you need to make sure that all the actors are fluent in their lines and confident about where they should be on stage throughout the piece.

If there are monologues, ask the actors to record the work on video (or make sure that everyone watches them) so that it is easy to point out the most effective moments. Give notes to the cast on voice, movement and facial expressions.

Consider also the **proxemics** and the relationships between the characters.

When you perform, your audience should be entertained and be aware of the point of view you have of your theme. Give your work a title and create a poster or simulate a brochure page advertising your devised work.

The flyer in Figure 3.11, for example, gives a clear idea of the theme and style of the piece. Is it an exuberant and funny parody of the **Western** genre.

> **Useful tip**
> When your work is in its final state make sure that you use your rehearsal time effectively.
>
> At the end of each session make notes about what you will rehearse next time. It is very tempting to start at the beginning each time. This will mean that certain scenes will be neglected and maybe that some people in the group will be very bored if you never get round to their scenes.

> **Key terms**
> **Proxemics:** describes how close or far away actors are from each other on stage and how this is used to convey dramatic meaning.
>
> **Western:** a genre in literature, drama and film which tells stories set in the American West during the late 1800s. Westerns often feature American Indians, the original inhabitants of the land, and cowboys, the new settlers.

Figure 3.11 Spittoon *is a parody and homage to the Western in all its forms*

3.6 Devising work using improvisation

If you were going to the theatre to see an improvised play, what would you expect to see? Would it be a series of funny quick-fire sketches or a play which was being made up on the spot? In fact, you would not be able to tell whether the starting point of the production was a script or a series of **improvisations**. By the time the play is performed in front of the audience, it will have been designed, costumed, lit and fully rehearsed.

You can use improvisational skills to:

- devise and develop any piece of work
- develop an understanding of a character in a text you are working on
- develop an understanding of the relationships between characters in a play you are working on
- try out scenes in a play you are going to see
- try out technical and design ideas.

You may start with a given task, for instance: to create and develop a character. Then, in a group, you can *structure* your material so that it will engage and interest your audience. You will also have the opportunity to make decisions about stage form, setting, costume and all other design and technical elements. When your play is fully rehearsed, it will be ready to present to your audience.

The renowned British film and theatre director Mike Leigh has devised many improvised plays, such as *Abigail's Party*, *Two Thousand Years* and, more recently, *Grief*. He begins by working with a team of actors who each create and develop a character. He then sets up scenes and selects the material he wants to include in his final play. This is a long and detailed process.

> **Key term**
>
> **Improvisation:** when actors invent or make up a script spontaneously. In devised theatre, this is a useful method to develop characters and situations in order to create a play.

> **Useful tip**
>
> When you are improvising, keep notes after each session and record targets for the next session. It is also useful to video work in progress.

Figure 3.12 Two Thousand Years, *National Theatre, London, UK*

Getting started

Activity

With a partner, try this series of short improvisations:

- a meeting between a parent and a teacher about a difficult student
- a young person explaining to a parent that they are leaving home
- an estate agent giving details to a client of a property that the agency is desperate to sell
- a business executive complaining about a late train to a member of the station staff.

When you have tried out these short improvisations, decide who you think is "in charge" in each scene. It may not be the obvious person.

Now pick one improvisation from those you have just tried. Try it again but, this time, put emphasis on the **status** of the character you think is "least in charge". Your aim is to raise their status. You can do this by thinking about your *body language*, *gestures*, and how you use your *voice*. And, of course, the things you say will have a bearing on the status of the character.

You can try a whole series of changes to develop the improvisation. For example, you could change the age of the characters or alter the mood of one character as the improvisation progresses.

Activity

Choose the scene which you think is working well in showing the characters, and **structure** it by deciding on:

- your starting point
- the set you will have on stage
- the props you could use.

Rehearse the scene and present your short piece of polished improvisation to your audience. You can show:

- the relationships between the characters
- the moments of conflict
- a clear narrative line.

Remember

You and your partner do not both need to be on stage all the time.

Key terms

Status: the relative position or standing of someone in a group to others, for example, the ship's captain has a higher status than the surgeon but lower status when ill.

Structure: the organisation of work, in terms of its starting point, its setting and narrative line.

Useful tip

When trying to show status, power isn't necessarily shown by shouting or towering over someone.

Devising practical work

3.7 Devising work using a character

All types of drama need characters, and improvised work is no exception. In scripted work you can rely on the playwright for the script, characters and theme, but in improvised work you must rely on your experience, imagination and research.

There are many different **stimuli** which can be used to create characters, but a useful starting point is to think of someone you know. Choose someone who is older than you and of the same sex – it's hard enough developing a character without having to act a different gender as well. Choose someone who is not known to other people in your drama group or your school. You need not know the person very well because you will use your imagination to build up details that you do not or could never know about them.

Building a physical picture

Try these individual exercises.

Key terms

Stimulus (plural stimuli): the starting-point for a devised work; the idea, image or object that sparks off your work.

Mime: using clear gestures and movements but no words to convey a character's personality and emotions.

Physicality: a character's unique physical features (presence, gestures, posture, etc.) which the actor will need to convey.

Activities

1. Imagine that you (as your character) are waiting:
   - in a doctor's waiting room
   - at a bus stop
   - outside the boss's office.
2. Try to get a sense of how your character would feel and what they would look like in each situation. Show how they would stand, any gestures they might use and anything they might have with them. There is no need to do anything 'dramatic'. The point is just to think about their **physicality**.
3. Now think about what your character's kitchen would be like and, in role, make a hot drink. Think all the time about what they would do, and **mime** the whole sequence.

    All the time you are working, show in as much detail as possible how you are doing everything: how you fill the kettle, open the cupboards, and wait for the kettle to boil. Build up a picture in your mind of the style of kitchen, the type of hot drink being prepared, and the general tidiness of the kitchen.

Extension activity

Imagine that your character is going out to meet with friends. Mime opening their wardrobe and looking through and selecting the clothes they might wear. This will give you the chance to think about their personal appearance.

- Do they spend lots of money on their clothes or are they creative on a small budget?
- Are there particular styles they favour?
- What about colour and fabrics?
- Are fashion trends important?
- Would they dress up for meeting friends or not bother?

Building a vocal picture

Try out different ideas to help you think about the voice of your character. Again, imagine your character in various situations where you can get a sense of the different tones of voice and the language they would use. Remember that everyone has particular ways of speaking which are appropriate at different times. We use different language with our friends than that we would use, for example, at work. In formal situations our choice of vocabulary is appropriate to the situation we find ourselves in and, if we are trying to persuade someone to do something for us, then not only our language but also our **tone** and **pace** changes.

> ### Activity
>
> Still working alone, use the idea of telephone calls to experiment with the voice of your character. Practise your character trying to make a call to:
>
> - persuade someone to give you a lift
> - make an excuse
> - make a complaint
> - arrange to meet someone.
>
> How does your character's voice change in each scene?

> ### Key terms
>
> **Tone:** using your voice to express what you are feeling.
>
> **Pace:** the speed and rhythm of your speech and how you pick up cues from others.

Figure 3.13 The Grizzled Skipper written by Maggie Nevill

Think about the *pace* of your speech:

- Will it be fast or slow?
- Thoughtful or hasty?
- Halting or lively?
- Measured or varied?
- And how loudly will you speak?
- Are you pausing to think or are you interrupting?

Think about the *tone* of your speech:

- Will it be shrill at times?
- Harsh or gentle?
- Aggressive or warm?
- And how appropriate is the language you are you using?

Already the character you are creating will have moved away from the real person you started with and your imagination will be filling in the gaps.

Devising practical work 41

3.8 Developing a character for devised work

You will now need to develop the role into a believable character who you can put into a play. One of the most straightforward ways of doing this is to work in small groups of four people and to **hot-seat** each character in turn.

You must answer the questions put to you *as the character* because this will give you further practice in sitting, speaking and moving as him or her. Now is a good time to invent a name for your character, but avoid choosing one that is too close to the name of the person you first started with.

There are hundreds of questions you could be asked and you will have to be quick-thinking and answer straight away, inventing answers and therefore details of your character. The questions put to you will be basic to start with, asking things such as name, age, marital status and employment status. Inquisitive members of your group might then move on to ask about what sort of car you drive or what sort of house you live in so that they can work out how much money you have. You may then be asked about what is important in your life so that they can find out about your beliefs and your personal relationships.

> **Key term**
>
> **Hot-seating:** the technique of an actor staying in role while answering questions from others in the group about the character's thoughts and feelings. The actor can involve the audience by asking them for advice.

Figure 3.14 Two actors rehearsing The Cage, *written by Deborah Gearing*

Remember, you should keep notes throughout the development process. They may be useful for when you are constructing your play and will be essential for when you are performing. You could record your notes in diary form, showing how you have progressed from each practical session.

Alternatively, you could make a poster of a profile page for a social networking site that includes basic factual details about yourself, including your employment, interests and friends. This will help to build up ideas of your character and their relationships with others. You could add a photo of yourself looking like the character, or draw a picture, or choose a photo from a newspaper or magazine – but avoid choosing anyone recognisable or famous.

Figure 3.15 Using online aids such as social networking sites can help in visualising new characters

Activities

To further develop your character, try these exercises.

1 Work with a partner. One of you will play the part of your character and the other will be in role as the parent.
   - The scene is at home, and the time is 6.00 pm.
   - The parent is at home and you, the character, enter.
   - You have some very good news to tell.

   Set the scene and run the improvisation. You will find out what actually is good news for your character and how s/he behaves in this situation. You may also be able to use it in your play. (Remember, what is good news for you may not be good news for the parent.)

2 Now you can swap over and try different variations and scenarios. You could:
   - give some bad news to a friend
   - share gossip with a work colleague
   - reveal a secret to a sibling
   - tell a lie to your boss.

   You may be able to think of many different variations to suit the character you are playing.

Devising practical work

3.9 Devising from a character

Communicating your character to an audience

You need to see if your character can be performed and appreciated by an audience, so you will need to develop, rehearse and perform a short piece.

Work by yourself and prepare a **monologue** to perform. You could write it and learn it if you wish. Assume that it is a speech which will be the opening moment of a play and, as you will be on stage alone, you could imagine you were making a phone call or answering a call. You could be recording a video diary or talking directly to the audience.

All drama needs **conflict**, and so in your speech show that you are struggling with something in your life. For example, it may be a relationship problem or it may be a decision you have to make. You should be able to give your audience a sense of where you are and who you are, as well as what is happening and the mood of the speech.

You will need to consider:

- Appropriate language – formal or conversational?
- Tone of voice – harsh or gentle?
- Facial expressions – anxious or angry?
- Gesture – fussy or controlled?
- Movement – agitated or confident?
- The relationship with whomever you are speaking to – friend or an adversary?

> **Key terms**
>
> **Monologue:** when a character on stage speaks alone, sometimes directly to the audience.
>
> **Conflict:** an element of struggle, found in all drama; it may involve trying to resolve a problem or someone changing their life; it does not necessarily mean an argument.

Figure 3.16 From a production of The Cage *by Deborah Gearing.*

When you perform your work, if possible, use costume, props and lighting to give a sense of place and time. Decide what mood you want to put across and use movements, gestures and voice so that the audience gets the message.

The following extract is one of a series of short monologues from *A Memory of Lizzie*, a fictional account of the childhood of American murderess Lizzie Borden. Here, Eliza is giving evidence. Although she is not alone on stage, notice how the stage directions make it seem as though she is, by isolating her.

> ### Activity
> Try performing this monologue. Work with a partner as a director and use your movement, gesture and voice to show how horrified Eliza is. This monologue could also be played from the perspective of a male character.

> **Useful tip**
> You should be noticing and noting down acting skills whenever you watch live theatre performances. Look for convincing moments in the work of others. If you are saddened or amused by what they do, note how they have had this effect on you. Was it what they said? How they said it? Their tone of voice? Their posture? Gesture? Or facial expression?

> *(Main stage lights dim, spotlight on Eliza. The rest of the cast is in the shadows.)*
>
> **Eliza:** I went into the sitting room and saw the form of Mr Borden lying on the sofa. His face was very badly cut with apparently a sharp object. There was blood all over his face; his face was covered in blood. I felt his pulse and satisfied myself he was dead, and took a glance about the room and saw that nothing was disturbed at all. The body was lying with its face to the right side, apparently at ease, as anyone would if they were asleep. I could hardly recognise the face.
>
> *(Stage lights back up, spotlight out. Break the freeze slowly.)*
>
> From 'A Memory of Lizzie' from *Sepia and Song* by David Foxton.

As well as performing your own monologue, you will also be a member of the audience for the other actors in your group. When you are watching, you should make notes on what the actor did. Keep your notes simple but be precise. Write down how the performance made you feel and what the actor did to achieve that effect. It may be that they spoke slowly and used a low tone to show their sadness, or that they sat slumped in their chair. They may have entered slowly, dragging their feet. On the other hand, they may have run in laughing as they did so and showed how excited they were by fidgeting with their hair.

What is the actor in Figure 3.16 conveying? He is alone on stage and speaking directly to the audience. Take note of:

- his costume
- the props
- the setting
- the lighting
- his physicality
- his facial expression.

Think about what he might be saying and how.

3.10 Devising work using a photograph

In the pre-release material for Cambridge IGCSE, you may have a photograph as a stimulus for devised work. These photographs tend to be documentary 'real-world' photographs. In this section, we will look at photos in this book to give you an idea of how to gather evidence and clues from photographs, and how to include your ideas in developing a piece of work.

Looking at the image in Figure 3.17, what can you tell about the characters?

- Who are they and what is their relationship to each other?
- Where they are?

Figure 3.17 From a production of Billy Liar *in 1960*

There is an older character and a younger character and there appears to be some conflict between them.

Look at the man at the back and his facial expression. Is he calling, explaining, shouting, singing? His stance is leaning forward and is focused on the younger man. He is tidily dressed in an old fashioned sweater and a neatly tied tie.

The younger man is unkempt in appearance – scruffy around the neck with wild looking hair. His attention is upwards. Is his facial expression bemused or surprised? Is he listening to the other man or commenting on something in the newspaper? He is wearing a coin in his right eye, which seems to represent a monocle.

They appear to be indoors somewhere and the period is the fairly recent past, suggested by the clothing and the newspaper.

Activity

To develop your ideas about what is happening and why:

- In pairs, re-create the moment: make a tableau; a still image.
- Decide on the relationship between the two.
- Think about the status of each character: who seems to have the upper hand here?
- Now go back to Section 3.8 'Developing a character for devised work' and go through the hot-seating exercise.
- Improvise the scene from the photo and take it in different directions. Swap roles and change the relationships until you find an idea that appeals to you.
- Rehearse and present your scene.

Using photographs

Photographs capture just a moment in time, and, as a stimulus, they can be interpreted very differently depending on whether they represent a starting point, a middle scene or the final moment of a play. Look at Figure 3.17 again, and imagine that is going to be the *final moment* of your play. Think about the journey each character has taken to get to this point.

Think about what genre you will create. Could it be:

- comedy
- tragedy
- mystery
- thriller
- musical?

Now turn back to the photograph at Figure 3.13 on page 41. This shows just one character, but there is lots of evidence about him from his facial expression, his posture, his haircut and clothes. To develop the character, you could write a script for what he is having a conversation about. Who is he speaking to?

You can try lots of different ideas, for instance:

- Is he hearing bad news?
- Is he listening to instructions?

Try this as the *starting point* of the same play.

Now turn back to the photo at Figure 3.14 on page 42, and, using the same technique, develop and investigate the characters and the situation they are in. Use this as the *middle scene* in the same play and work out how they got to this moment. You could develop friends and family for both the characters and structure scenes leading to this moment.

Ask yourself: *what will happen next?*

Extension activity

Turn back to the photograph at Figure 3.12 on page 38. This shows a family in conflict.

- Each take a character, make a tableau of the moment and think about how you are feeling and who you are relating to on stage. Are you listening or speaking? Where do you think the focus of the scene is?
- Now discuss the relationships, the location, and the situation the characters are in, then try the tableau again.

Useful tip

Make it clear where you are, who you are and who you are talking to. You can introduce as many other characters as you wish to develop and structure the piece.

Devising practical work

3.11 Making your play

When you start to put ideas together for your play, you will need to be flexible and creative. When your group comes together you will all have lots of information about your own character and that of others. A theme or storyline may have begun to suggest itself to you, or you may have been inspired by some preparatory work you have done.

If you have worked on monologues you may have found one really interesting and you may want to use it in your performance. If so, it may become the starting point for your work, or it might be the conclusion. You could use it as the cue for a **flashback** moment.

You may need to make some adjustments to your characterisation so that you and your group can develop ideas about how all the characters can come together in one piece of work. You may, for example, need to make your character much older, but it will be interesting to try out improvisations with your character at different moments in their lives.

You may decide to make all the characters much older than at first.

Structuring your play

In your play, make sure that each actor has a specific and different role so that you do not end up with characters with the same point of view, as this can lead to repetitive work. Make sure that your storyline is clear and that all members of the group have the opportunity to show their skills. Be sure about the journey each character will take and any conflict within or between the characters.

It's a good idea now to write a **scenario** so that you get an idea of the narrative and also of the shape of the work. You can use a variety of scenes so that the audience will remain interested throughout. You can keep a list or make a series of sketches or photographs to keep track of what's happening.

You will need to make some decisions about the period that your work takes place in. You have probably set it in modern times but there may be scope for varying this. This decision will have implications for costume and set design.

Look back at the advice on page 31 and look also at Viola's speech in Twelfth Night (on page 16). Here is how it fits into the structure of the play, with the number of characters in each scene shown in square brackets.

<u>Act 1 sc. 5</u> is a busy scene in which Viola delivers Orsino's message of love to Olivia and is sent away. Malvolio is to follow Viola to return a ring which Olivia pretends Viola has given to her. [5]

Olivia and her household are introduced and the theme of unrequited love is established.

<u>Act 2 sc. 1</u> is the following scene which introduces two new characters, Antonio and Sebastian, and is a short conversation between the two. [2]

The scene lets the audience know that Viola's brother is alive, thus developing the plot.

Key terms

Flashback: when the narrative in a drama switches from the current time (current from the point of view of the characters) to an incident from the past, perhaps as a memory or a dream by one of the characters.

Scenario: the summary or outline of the plot of the play and a list of the scenes in order and the characters which appear in them.

Activity

Try improvising the following situations:

- at a school reunion, looking back at your younger days and showing how well you've done
- being interviewed for a promotion at work, perhaps by someone you used to know
- at a retirement party, looking forward or maybe back.

Remember

Don't feel that all group members need to include their monologues, as this can make the structure of the play rather predictable.

Act 2 sc. 2 is where Malvolio catches up with Viola. After a short conversation, we have Viola's soliloquy. [2, then 1]

This scene complicates the plot and tells the audience Viola's feelings.

Act 2 sc. 3 is a scene of comedy, singing and plotting. [5]

Maria sets up the trick that will be played on Malvolio.

Shakespeare structures this part of the play to develop the plot, introduce new characters, develop the sub-plot and vary the moods thus keeping the audience intrigued, amused and interested in the action. Note also the varied number of characters in the scenes.

Rehearsing and performing

Now that your play is set and no further changes will take place, you can concentrate on the performance of your role.

While you are rehearsing you will be refining your performance skills and working on your movement, gesture and voice and, of course, on how you will react to other characters on stage. The way you use your acting skills will make it clear how you feel about the other characters as well as what you actually say. Check the following:

- Are you using vocabulary appropriate to your role? (This may mean that you will have to do some research, especially if the character does a job that you are not familiar with.)
- Is your **register** appropriate? (e.g. if you are playing someone older than you, avoid using slang expressions)
- Is your accent appropriate?
- Is your tone of voice different when speaking to different characters? What does that show about the relationships?

How does your physicality show:

- whether the person you are speaking to is superior or inferior to you?
- whether you like them or not?
- whether you want them to do something for you?
- whether you are afraid of them or they are afraid of you?

As you approach the technical rehearsal, make decisions about set, costume, props, lighting, and any other technical devices. At the technical rehearsal, you should go through your play from **cue to cue** and finalise all set, costume and lighting changes. Don't worry about the acting, apart from the entrances and exits.

Once this is set and recorded, you will be ready for your **dress rehearsal**. This should be as close to the actual performance as possible. Try to have an audience, and pay attention to their feedback. Simple points can be very helpful, such as whether they can hear you and whether they can actually understand what is happening, as well as more detailed information about what they noticed about your acting skills.

If you are thoroughly prepared you will really enjoy your performance, and so will your audience.

> **Useful tip**
>
> When you get the pre-released materials for Component 1, look back at the processes you have already used.

> **Key terms**
>
> **Register:** the tone or formality of language used to convey a particular character or setting.
>
> **Cue to cue:** go through the play to all moments when there is any technical change (to lighting, sound or set) and rehearse them.
>
> **Dress rehearsal:** a full rehearsal (as of a play) in costume and with stage properties shortly before the first performance.

Devising practical work

3.12 Devising using a specific audience

It is always important to consider the needs of your audience when you are planning any production. In this section, we will look at devising a piece with a particular audience in mind. You may already have a **target audience** in mind. This may be your own age group, parents and adults, or perhaps younger students at a local junior school. In each case you will need to consider carefully the style you take, the theme you choose and the language that will be most appropriate to the occasion. These thoughts will then become a part of your **directorial concept**.

Research your audience

Once you have decided on your target audience, you will need to know the topics, issues or themes that will be most appropriate or relevant to them. If, for example, you have chosen to perform to new entrants to your school on an induction day, you will need to first gain permission to present your piece to them and find out how long the performance should last. The next step is to gain permission to use the performance space you wish to use and to make an official booking. Once you have booked the space and declared that you will perform, you have made a commitment – there is no turning back – the show *must* go on!

Research your theme

Whatever ideas you have, it is vital that you research them thoroughly so that you can be confident you know what you are talking about. It will also give you greater confidence in your performance. Since performing for new students has been suggested as an audience, 'starting at a new school' could be an appropriate theme for them, so this is the example we'll take.

Changing schools always raises hopes and fears. Ask the teacher in charge of the first year in your school what problems are frequently faced by new students and what is done to help solve them. Perhaps you could also contact their previous teachers for advice, but make sure you have planned your questions carefully – these are busy people. Some of your group may have younger brothers or sisters who can help in your research, and don't forget that you were once new students yourselves. Draw upon and share your experiences, not forgetting to make clear notes. Starting school is an important part of growing up and you will certainly find useful material in biographies, novels and plays.

Research your materials

During your research, you may find written material that is relevant and this could help you decide on the direction you take in devising your production. A good start is the play *Daisy Pulls it Off*, by Denise Deegan.

> **Key terms**
>
> **Target audience:** the audience for which the production is specifically devised.
>
> **Directorial concept:** the basic ideas that underpin your production and that you want to communicate to the audience.

In the first act, the heroine arrives at her new boarding school and encounters the school bullies. The period is the 1920s and the style is deliberately exaggerated with period slang and direct addresses to the audience (see Figure 3.18). This style contrasts with that of the novel and play, *Kes!* by Barry Hines. Set in a secondary school in the 1960s, it is more naturalistic in style and language. Both offer set pieces such as confrontations with authority, bullies and the games field that could be set against each other for contrast. Among these you would be able to knit together material drawn from your own experiences and shaped to fit your production. Though the style is not realistic, audiences can recognise the emotions as being true to life.

So far we have looked at the topic of starting a new school from the point of view of the pupils. Remember that in every situation there are other viewpoints. Think about how the parents might feel, or imagine what it might be like to be the teacher in charge faced with coping with a new intake. Whatever theme or topic you choose, you will need to consider and include a range of different viewpoints in order to show **balance** in your production.

Useful tips
- Find out the needs of your target audience.
- Write them down on posters placed in your drama room.
- Keep referring to them to keep on track.
- Consult frequently with your drama teacher.

Key term
Balance: giving fair attention to other viewpoints so that the production is seen to be unbiased. This is very important when dealing with controversial topics.

Useful tip
It might be helpful to look back at earlier sections to remind you of the practical skills you can use.

Figure 3.18 *Daisy Pulls it Off performed at the Lyric Theatre, London, UK, 2002*

Activity
- Collect a range of written materials on your chosen theme, e.g. starting at a new school.
- Choose short extracts and organise them into a possible running order.
- Improvise ways of linking between the extracts.
- Perform your programme based on a theme.

Devising practical work

3.13 Planning the production

Work as a group

It is best to work co-operatively, with everyone contributing to the devising and writing of the production. This means that everyone can have a say in the way the piece is formed. It is always important for the design and technical contributors to be closely involved in the planning. Contributions such as offering ideas for costume, set or lighting, can help shape the way the show develops.

Consider involving your audience in the action

It is important to always to consider your audience when preparing your production, and whether you want them to be observers or to be more involved in the drama. **Forum theatre** is an example of performance in which the audience is very involved. There are a number of factors which might influence the involvement of the audience in your play. If your performance space is small and intimate, and your stage is not raised, you may have a closer connection with the audience. Lighting and sound can also be used to involve the audience to a greater or lesser degree.

One of the simplest approaches to involving your audience is to give them a role in the production. You might wish to explain to them before the show that they will be treated as a jury, for example, and will have to vote on a decision at the end of the show. More subtly, they can be greeted at the start of the show by an actor **in role** who welcomes them, for example, as members of the staff or school council and explains that they will have to make a decision based on what they are about to see. See if you can build in moments of **hot-seating** or **forum theatre** that will work as part of the play. The approach you choose will depend on your confidence in yourselves as performers and your confidence in the audience's ability to accept you. Careful preparation will certainly help build your confidence.

Involve your audience's sympathy

If you choose the starting school topic from the previous page, you might wish to present your show through a character that the audience might sympathise with. Perhaps the production could be centred on someone who is frightened at the thought of going into secondary school and who shows all the anxieties that were revealed from your audience research. If the character seems more anxious than members of the audience, it is likely that they will be sympathetic. The character could address them and ask for advice at key moments and help them feel involved.

In rehearsals, members of the group who are not performing could usefully take on the role of audience and ask questions or give advice. It also helps the performers as they can be prepared for different kinds of response from their audience.

> **Key terms**
>
> **In role:** appearing convincingly and consistently as a character different from one's self.
>
> **Forum theatre:** an interactive form of theatre developed by Brazilian director Augusto Boal. The audience stop the play to suggest different solutions to a problem that a main character is experiencing.

> **Remember**
>
> - Don't be over-ambitious. Only try approaches that you and your group are comfortable with.
> - Consult with your drama teacher and those who know your audience well. Take notice of their advice.

Experiment with hot-seating

Hot-seating, or having an actor in role answering questions from the audience, can be a very effective way of involving your audience. You may already be experienced in the technique from your work in improvisation, in which case it should not be too difficult to apply it in live performance. If not, use it as a tool in rehearsals to help you create a rounded **characterisation**. Don't consider using the technique until you feel absolutely confident that you can respond comfortably as your character in all situations.

Hot-seating can be used effectively at the end of a performance by giving an audience a chance to ask characters about their **motivation.** During the show, hot-seating can help an audience to understand why something is happening. Experiment in the safety of rehearsals and see what ideas come out. Follow leads and be flexible – you can do it if you try!

> **Useful tips**
> Remember to keep notes of your planning process:
> - what you did
> - how you did it
> - why you did it.

> **Key terms**
> **Characterisation:** the way in which an actor presents a character in a play.
> **Motivation:** reasons why a character does something or behaves in a certain way.

Figure 3.19 Students from Beijing National Day School performing a drama based on real school experiences

Activity
- Choose a character you have played previously in a scripted play or improvisation.
- Think of ten questions you would like to ask the character that were not answered in the piece you performed.
- Write down what you think your character would have replied to each question.
- Try answering questions about your character from members of your group. Start simply until you become confident.

> **Useful tip**
> Go back to Section 3.8 for further help on developing a character.

3.14 Preparing for the show

Organising your show

Careful and detailed planning is essential. You will have to research your audience and materials to fit and support your chosen theme or topic and you will have to consider ways in which to involve your audience. However, directly involving the audience can throw up unexpected problems, so it is safest to be prepared. When in doubt ask your teacher for help and advice.

Plan your audience management

First, you will need to consider how your audience might behave, and you will certainly need to consider how best to manage your audience. Even the most experienced performer can 'wind up' an audience to the point at which they are all calling out and not listening. It takes skill to get them quiet and attentive again. Throwing sweets into the audience may be fun at a pantomime but may be less so if you are wanting the audience to listen to an important point you have to make in your show.

You will need to plan how you will manage the audience when it arrives. Traditionally, in a theatre, the audience arrives and you perform. In this style of drama there are many more possibilities open to you.

It may be possible to arrange the audience into smaller groups, each led into the performance space by an actor who can talk to them informally but still in role. The actor can then act as leader and focus for the group, feeding them ideas and offering particular viewpoints. Smaller numbers mean that the group can feel more confident about contributing ideas. Also, each group can have slightly different viewpoints, which will make it easier for having discussions at the end if that is how you wish to end your show.

Figure 3.20 An actor interacts with his audience

Choose your performance space to fit the show

Think about the performance space you use. This could be led by the needs of the show, for example, in a playground if your audience research into "starting school" suggested that most of the audience was anxious about being bullied on their first day. The audience could be led by sympathetic characters into the playground where the rest of the cast are already performing as bullies. The action could be frozen by a character playing the teacher blowing a whistle or ringing a bell. It is always a good idea to make sure that the roles of characters that have to lead the audience or give instructions have a high **status** so that what they say will be listened to and acted upon. Actors could then move onto the performance space and perform their parts until the next whistle.

Whatever performance space you choose, think carefully about practical issues:

- Is there sufficient space to accommodate the audience comfortably and safely (there may be health and safety regulations that you need to comply with)?
- Will everyone be able to see all of the action?
- Are there opportunities for interesting levels on stage?
- Is there scope for having a variety of entrances and exits?
- Is there sufficient space backstage for actors and crew?

Figure 3.21 *Students from DPS International School, Saket, India, performing a devised piece*

Useful tip
In your planning, make sure that you share out the parts fairly and that everyone has sufficient exposure on stage.

Key term
Status: the relative position or standing of someone in a group to others, for example, the ship's captain has a higher status than the surgeon but lower status when ill.

Activity
- Make a list of the tasks your group will need to do in preparation for your piece.
- Number the tasks in order of when they will need to be done.
- Against each activity, write the names of the members of your group who you think will best complete that task. Don't use the same name twice!
- Plan how you will complete your assigned task.
- Draw up a tick list to help you check when you have each completed your tasks.

3.15 Bringing your show together

Turn your research into a production

So far we have looked at the things that make productions with the audience as a stimulus different from other performance options. Now we will look at putting material together into a show. As this unit has focused on devised work, you will already know how extracts from play-scripts, poems or prose can be fitted together with original work that you and your group have produced. When developing a theme for a specific production, you can just use original work that you and your group have improvised or written if you wish.

Devising means making a commitment to your audience and the theme that will be appropriate for them. Your group will need to agree firmly on the theme. Ask yourselves the questions:

- Why have you chosen the topic?
- Why is it important, both to your audience and you?

If you can answer these questions confidently then you are ready to devise.

Have a group meeting where everyone involved shares their research and ideas. Write these down on large sheets of paper and pin them up on the wall. Spider diagrams, sketches, photocopies of play extracts, and photographs can all be pinned up with notes in different coloured marker pens to help keep track of ideas. These can also show the contributions to sound, lighting and design.

- Always keep your target audience in mind.
- Think about **audience participation** so that you can involve them directly in the show.
- Remember that you have a responsibility to your audience – if they are apprehensive about starting a new school then you will need to focus on ways that will reassure them.

Experiment through improvisation

Earlier, we chose "starting school" as an example because it is a topic that is very familiar. The way in which we looked at that idea is the same for any topic. Out of the research will come ideas. These can be developed through further research as you add to or replace written materials. *Tom Brown's School Days* by Thomas Hughes, *Nicholas Nickleby* by Charles Dickens, or Roald Dahl's *Boy* might give material on bullying or unsympathetic headmasters that could be quoted, acted out or used as a starting point for improvisations.

The ideas raised from talking to people or remembering you own experiences of starting school, for example, will give you very useful starting points for improvisations. The activities box will give you some starter ideas.

Useful tip

To see how a theatre company might go through the collaboration process, watch the videos on Devising at the National Theatre website.

Link

www.nationaltheatre.org.uk/backstage/devising

Key term

Audience participation: directly involving the audience in the production, for example, by asking them questions or giving individuals simple tasks.

> **Activity**
>
> Improvise scenes involving a new student and a parent:
>
> - In a shoe shop buying uniform shoes – the new student insists on something fashionable; the parent doesn't want to buy them.
> - Five a.m. on the first day – the student is in uniform, the parent is in bed…
> - Breakfast time, next day. The student starts with "I don't feel well enough to go to school…"

Shape your material

Once you have collected the materials, you will need to give them a structure so that they make sense to the audience and fit together in a satisfying way. Basically, it needs a beginning, middle and end. In the case of "starting school" you will want your audience to leave happy, reassured and wanting to come back at the start of the school year, so the end might be the section you think about first. Perhaps you would devise a cheerful, up-beat song and dance number in which members of the cast take groups away to meet friendly students and discuss what they have seen. Consider the ways in which you can **pace** the drama so that there are moments of action and reflection. The worst approach is to start at the beginning and rehearse the opening over and over until you realise it's two days from the show and you still have not done the ending.

Consider beginnings and ends

Carefully plan your opening scene and grab the audience's interest and attention. Whether you have a shock or comic start, make sure that the next scene is different in tone, pace and mood. Earlier we looked at audience management. Remember that the structure of your production can help by following exciting actions with quiet, calming, moments.

> **Key term**
>
> **Pace:** rate of speed or movement of action or speech.

> **Useful tip**
>
> For more on pace, look back to Unit 2, page 18.

> **Link**
>
> www.upstarttheatre.com

> **Useful tip**
>
> Look at earlier sections for useful approaches. Look at the website of Upstart Theatre Company in Perth, Australia to see how a professional company works at devising a show for a specific audience.

Figure 3.22 A Disappearing Number by Complicite, directed and conceived by Simon McBurney (2007)

Devising practical work

3.16 Performance styles

You have been learning how to prepare for, and perform, a piece of theatre based on a theme designed for a specific target audience. The audience could be of any age but you must recognise and cater for the specific needs of your audience and choose approaches that will make the impact of your work most effective. These are aspects that have been covered in this unit so far.

The skills you will need

We have already looked at performance skills in Section 3.6 and these will support you in any project you undertake – they are transferable. As you develop as a performer, you will find that it will become easier to think of both what you want to say and how you will communicate it. Drama is about giving you a voice – we hope that this book helps you find it.

Experiment with style

Before we start, we need to know about **genre** and **style** and how they are linked (see Unit 2 for more on genre and style).

- "Genre" refers to the type of production, for example: comedy, tragedy or documentary.

- "Style" is used to describe the way in which the production is presented and performed, such as in a **naturalistic** or **expressionistic** way.

Each production will require a different approach and even as you work on a particular show you can usefully experiment with style. Once you have a clear idea of the shape of the piece and the material you are going to use, you can start to rehearse using different approaches. Here are some examples of genres that you may wish to consider. For each of these there are **theatrical conventions** as to how they are performed and presented.

Documentary

Documentary genre involves presenting the material in a balanced, factual way as if in a serious television programme. This is particularly appropriate for topics based on real-life incidents, perhaps featuring real people. Dramatised flashbacks might be presented in a serious, naturalistic style, for example, in *A Memory of Lizzie* in *Sepia and Song* by David Foxton.

Pantomime

You might be able to present your material in the genre of a pantomime using stock characters such as the Fairy Godmother (or play with the names to produce the Hairy Godfather for instance, as a mafia-type villain). Here, instead of naturalism, the characters would be presented as caricatures with broad gestures. They might follow the traditional pantomime conventions of a principal boy played by a female and the dame played by a male.

Key terms

Naturalistic: attempts to faithfully represent real life on stage.

Expressionistic: drama that tries to show emotions rather than reality. The term also applies to other art forms and is often contrasted with "naturalism" or "realism".

Theatrical conventions: a set of rules or techniques that are particular to a specific performance genre, and help both to govern how the play should be performed and how the audience identifies the genre of the performance. These can include what the character wears and their gestures and way of speaking, as well as techniques such as flashbacks and soliloquys.

Did you know?

Pantomime is not to be confused with the art of mime, but is a traditional genre in England and other English-speaking countries. It is a type of musical comedy, often based on fairy tales, and usually performed between December and February.

Melodrama

Melodrama is a variation on pantomime, based on an 18th century theatre genre that evolved when British theatres could only perform plays if they had a Royal Warrant. Unlicensed theatres got around the law by performing their plays with musical backing. These plays often had clearly defined heroines and villains each with their own theme music and using an over-dramatic and exaggerated acting style with spoken asides to the audience. A good example for you to look at is *Maria Marten,* or *Murder in the Red Barn.* Though based on a real event, unlike a documentary, the melodrama is usually performed with exaggerated gesture and vocal delivery that is both artificial and comic, despite its tragic theme.

> **Remember**
> Whatever style you finally decide upon, make sure it is appropriate to your piece and your audience. Above all, be consistent and enjoy the closer relationship you will have with your audience.

> **Useful tip**
> You may wish to look at a short video of contemporary writers talking about style on the National Theatre website.

> **Link**
> www.nationaltheatre.org.uk/video/finding-a-writing-style

Figure 3.23 Melodrama *Maria Marten performed at the Arts Theatre Club in London, UK, 1942*

Finally

Whatever approach you take to devising, and whatever style and genre you decide upon, always make sure that it is appropriate to your theme. Above all, be committed to your work and enjoy the closer relationship you will have with your audience. We hope that you will have learnt new ways of devising and are enjoying taking part in the creativity involved in the devising processes.

Devising practical work

Unit 4 Design and technical work

There are many ways in which a simple performance can be made even more effective by using design and technical aspects, such as costume or sound. In this unit, we will look at the range available and see what each has to offer.

> ## Objectives
> In this unit you will learn to:
> - consider the range of design and technical aspects available
> - explore ways in which they could enhance a production

Even if you lack a stage and have no access to lighting or scenery, it is important that you consider the contribution they could make to your productions. Wearing a costume or a mask, for example, can help an actor to get into character, gain confidence and perform more compellingly. A well-designed set can communicate place to an audience, while lighting and sound can give an immediate change of atmosphere or mood.

In examination, you may answer questions that test your design skills based on your theoretical knowledge, **not** your ability to make the designs. However, opportunities to make your designs may enhance your understanding of the design process and provide valuable experience.

Most of all, the authors hope that this unit will give you an opportunity to enjoy learning about and gaining some of the variety of practical skills required in this area of the world of theatre.

This unit explores:

- costume design
- lighting
- make-up
- masks
- properties
- puppets
- set design
- sound.

4.1 Costume design

Costume means the clothes and accessories that actors wear in order to communicate their character in performance. Costume design offers you many opportunities to be creative and to help you in developing your role as a performer. Your costume can, for example, set your character in a specific period of time or setting.

Generating costume ideas

As a costume designer for a devised work, you will be involved in the choice of the theme your group will work on, so make sure that you keep a file of images that suggest the theme. You may need to research a period or style and your ideas may form the stimulus for performance work.

Costume can help to set the style of your performance piece and set it clearly in a specific time period. A character's status within the play is identified by the clothes they wear, and your audience will recognise this.

You will have some idea of what you want your design to say, but remember also that it will need to be practical. It will have to fit the performer and be comfortable. Make sure also that it meets the needs of what the performer will have to do on stage and that it is durable enough to last the run.

When we meet someone for the first time, we make an immediate judgment about them (whether we should do or not) and this is based on what they look like and what clothes they are wearing. Look around at the people in your drama group. Even if you are all wearing school uniform, you will all be wearing it in your own way and showing hints of your own personality. Out of school, your clothes say something about you: this is also true of characters on stage. Your costume design will be giving the audience clues about the characters they see and will also help the actors to feel more like the characters they are playing.

Some scripted drama needs costume to make a clear statement as to the characters you will meet during the play. Costume, make-up and masks (see Sections 4.3 and 4.4) can combine to create a wonderful effect and enhance the performance outcomes.

> **Useful tip**
>
> Find out what your costume budget is and research where you might buy fabrics to fit your budget, e.g. markets, fabric shops or even online.

Figure 4.1 Costumes can be used to show location, period and style

Design and technical work

Planning your work

Performers will often have to play more than one role or show more than one skill. As a costume designer you will be able to help the performance considerably by creating or designing costumes which will allow for quick changes and will capture the character perfectly to make the role seem more convincing.

You may decide to design a basic costume to be worn by all performers. Then you will be able to design additions for each actor to use to show their different roles.

In *Oh! What a Lovely War* (Theatre Workshop), the horrific story of the First World War is told through an **end-of-the-pier show** of that period and uses songs of the time.

Look at the photograph from a production of *Oh! What a Lovely War* in Figure 4.2. The characters are all wearing pierrot costumes in black and white with ruffles and pom-poms. The costumes suggest a uniform and are worn throughout the play. The different items, such as hats and other pieces of uniform, are added during the play to set the scene and differentiate the characters as the actors play a variety of parts.

Figure 4.2 From Oh! What a Lovely War *(Theatre Workshop)*

At the technical rehearsal, make sure that your performers know exactly where their costumes will be for any changes and that they have rehearsed their changes. Be there to help if they need it. At the dress rehearsal, take photos of your work in action.

> **Key term**
>
> **end-of-the-pier show:** In the late 18th century, the beach became a popular place of leisure for the British aristocracy. Many of these beaches had no harbours so piers were built to accommodate visitors: these are huge, metal structures that stretch out into the sea and are designed to allow travellers to step from ship to shore without getting their feet wet. The appeal of the pier soon spread and a trip to the seaside became a treat for working class. The end-of-the-pier performance developed to entertain this audience, in this unique setting.

Activities

Try the following to come up with interesting costume ideas:

- Look at your notes from productions you have seen and pick out any good ideas you would like to use.
- Come up with some abstract ideas which suggest the theme, for example, if your theme is money, design outrageous costumes showing excess.
- Come up with character-led ideas, for example, Cinderella's ugly sisters could be an inspiration for catwalk designs.
- Collect a variety of fabrics but do not limit yourself to traditional fabrics, for example, you could make designs using string, plastic, painted canvas, egg boxes, etc.
- Get hold of some old hats and remodel them suggesting the theme.

Useful tip

Make notes of any alterations you need to make.

Remember

When constructing your costume, you must remember that it needs to last for the duration of the show.

Costume design for scripted work

The costume designer has to consider a number of factors when working on a production.

In drama, 'costume' can refer to the style of dress of a particular place or time and for a particular activity. It can also include accessories. Costume will always have a huge impact on the audience, as the actor is transformed to that time and place as soon as they put on their costume, taking the audience with them. The costume influences the way the actor moves around the performance space.

Remember

The costume you make must be comfortable for the actor and be easy to get on and off.

Figure 4.3 The court scene in The Crucible *– note Mary Warren's costume*

Sometimes a playwright will give you clues in the script to show you what your character should be wearing. By finding these clues, your character development will be enhanced. *The Crucible* offers a range of challenges when undertaking the role of costume designer.

In Puritan communities during the period of the play (the 1690s), girls had to cover their hair and wear dresses that covered their arms and came down to their ankles.

Figure 4.4 The Crucible, *Royal Shakespeare Company, Stratford*

Activity

Sketch designs for a character in *The Crucible* or a play of your choice. If the character appears in different scenes, consider small changes to the costume. For example, how can you show the character is outdoors?

As the work develops

- Make a list of the characters and the sorts of people they are.
- Join in the hot-seating and ask detailed questions about the clothes they might like.
- Note if anyone needs particular accessories.
- Note if anyone needs a special costume, such as a secret pocket.
- Make notes about the relationships between the characters: you can give clues to the audience about these relationships using colour or style.

During rehearsals

You will be able to decide which costume you would like to make or assemble, or which you would like to provide sketches for. When you are putting your ideas together, you can use dress pattern catalogues and/or mail order catalogues for inspiration. Try going into stores or departments you wouldn't usually shop in, especially if you are designing for an older character. Fashion magazines are not particularly helpful as not many people dress in high fashion, even if they'd like to.

Improvised drama often requires contemporary clothes. Try talking to the actors to find out which shop you think your character might shop in, but be careful of actors who want to bring in their own best clothes and 'dress up'. Try to keep some control, even over the costumes you may not be designing for.

Remember to discuss your work with any other members of your group working on design so that you can think about colour scheme and, for example, the effect of lighting on your costumes.

Useful tip

When working on a script set in another time period, it is important to research the details of what people wore, including fashions, style, fabrics and colours.

In *The Crucible*, John Proctor's clothes are made to last and must be for working hard in. You may find items of clothing already constructed and adjust them to fit the period of time when the play is set. You must show the type of character he is through your costume design. What accessories would he need? A hat? What type?

Figure 4.5 *John Proctor's costume*

The Abigail Williams costume will be made from homespun woollen fabric and her hair will be covered in public, reflecting the strict Puritan society in which she lives. All the girls' costumes will also reflect these ideals, and they will appear like a uniform in some ways.

In the written examination, you may be expected to discuss how your use of costume contributed to the effectiveness of your performance piece. Wearing an appropriate costume will help you in the creation of your character.

Turn back to Figure 3.11 to see an example of how costume can be used to good effect in publicity materials.

Figure 4.6 *Abigail Williams's costume*

Useful tip
Keep a photographic record of your designs as well as your sketches to help with revision you may need to do.

Remember
- Don't forget practical details such as making sure that there is time for measuring and fitting your costume, budgeting and shopping for fabric.
- Collect accessories and be willing to help with hairstyles and make-up.
- Take note of health and safety matters; make sure that the costumes are fireproof and that they are comfortable for the actors to wear.

Design and technical work

Effective costume design

If your performance needs a historically accurate costume, you will need to research the period and location of the play. This could involve looking at pictures or paintings from that era. You will also need to consider how any accessories you wear – for example, a large belt, a straw hat and boots – would affect the way you move around the stage.

Costume can enhance the relationships between characters by showing visually that they are part of the same family or group. As a designer, you will have to interpret the characters in your production and find ways of showing an audience who they are and what they are like through the costume you have assembled.

For example, if you were to design the costumes for *Romeo and Juliet* by William Shakespeare, your choice of costume and colour would be used to immediately identify which characters are the Montagues and which are the Capulets. This visual prompt informs the audience before the characters speak.

Research the character and consider materials

The group discussions about style will give you a start, and you will need to discuss with the actors how the character will be played. Plan a colour scheme and think about fabrics and textures that will support your final ideas. As a designer you can help influence the acting performance, for example, by giving the heroine a pale and lightweight fabric that floats on the air as she moves, designed to contrast with the dark, rough textures of the villain. It is important that you understand the character fully, so if you can, improvise or read the part you are creating for in rehearsals to see what it feels like to play the character. Discuss your ideas with the set designer to make sure that colours and styles work with each other. Consult with the lighting designer and check the effect the lights have on your chosen fabrics – a smart red cloak will look black under green lighting.

Research the period

If your production is set in a particular period, then you will need to research the clothes worn at that time. Look at books on the history of costume, art books, photographs, paintings and postcards, as well as on the internet. Keep details of your research.

Consider the style of your production. Is it naturalistic? If so, you may wish to have your costumes as historically accurate as you can manage. Otherwise, you may be able to indicate period through details of **accessories** such as hats, or through the overall shape of the costume. The main things to consider are whether it effectively communicates what you want it to and whether it also looks right with the other design elements.

Experiment through sketches

Experiment with ideas by drawing sketches: it is the designer's particular form of improvisation. Keep them and work them into finished designs, adding swatches of fabric to give others a clear image of what you intend.

> **Useful tip**
>
> When designing a costume, consider the character's age, social class, job, wealth, interests and even nationality.

> **Key term**
>
> **Accessories:** items of clothing such as hats, belts, ties or jewellery that add to the overall effect of a costume.

Make or assemble

It is very useful to have had practical experience of making or **assembling** one of your designs. This will help you consider the practicalities. You will need to:

- take the measurements of the actor
- obtain the appropriate fabrics
- find suitable patterns
- sew and fit the garment.

If you are not sewing the costume, you will need to:

- source the individual parts of the costume
- assemble them on the actor.

Figure 4.7 Costume design for *Looking for JJ*

Figure 4.8 Researching costume design is integral to your production

Key term

Assemble: to put together items of costume that you have found, to make a complete outfit.

Remember

- Your costume must be strong enough to last for the run of the show.
- Think about health and safety issues – avoid trailing hems or heels that could cause an actor to trip.
- Photographs can remind the actor how the costume should be worn.

Did you know?

In early black-and-white Western movies, the audience knew who the bad guys were as they wore black hats while the 'goodies' wore pale colours. This is an example of a design convention.

Activity

Look at Figure 4.7. What do the costumes tell you about the characters? Discuss with your classmates.

Design and technical work 67

4.2 Lighting

Whether you are working on a scripted piece or a devised piece of work, lighting can make a huge difference to the effectiveness of the performance. The first thing to remember is that lighting is the provision of any source of light to enable the audience to see the action. It also helps to create mood and atmosphere.

If you have access to lighting equipment, then you will be able to experiment practically with different approaches. If not, then read on and consider some possibilities for giving the production you are working on a bigger impact on your audience.

You will need to research colour and design fully, as well as think in terms of focus and intensity. Consider how different colours can influence different moods: blues suggest cold and misery, while pinks give a sense of warmth and well-being. Red lighting is often associated with evil, and green often suggests a forest. Lighting can also be used to display costumes and to accentuate their colours, but make sure you always consult with costume and set designers, and experiment with colour effects first as lighting can also ruin the effect of a costume. Under green light, for example, handsome red costumes turn to a muddy shade of brown or even black.

Explore the different types of lantern you might wish to use. Flood lights are large powerful lights that can illuminate a whole stage, while spotlights are able to focus light intensely on a particular spot, drawing the audience's attention to a particular actor or part of the stage. Different lanterns are built for different jobs, so research **profiles**, **fresnels** and **parcans**.

Whatever you choose, remember that your lights must cover the areas where actors are working – beware of accidental areas of dark and shadow. Every lantern must have a purpose, so know why you are using it and what effect you wish to create.

> **Key terms**
>
> **Profile:** a focusable lantern with an ellipsoidal lens which enables a sharp beam of light to be projected. Also known as an ellipsoidal reflector spotlight (ERS).
>
> **Fresnel:** (pronounced fre-nel) the most common type of lantern used on stage, it has a textured lens and produces a very even light that is soft at the edges and tends to project a soft shadow.
>
> **Parcan:** a lantern that is used to provide strong dramatic keylight, backlight or effects such as beams of light in smoke.

> **Useful tip**
>
> Ensure your lighting is sufficient for all the audience to see the production. Audiences get tired easily if watching performances in semi-darkness for too long.

Figure 4.9 Opening lighting of the court scene in the Royal Shakespeare Company production of The Crucible

Scripted

If you are considering lighting for a scripted play, you will need to think about the most appropriate approach. Read the text carefully and look for clues – not only in the stage directions but also in references made by characters. If it is to be naturalistic, consider where the light would be coming from (light source). Different times of day and the weather produce different colours, intensities and directions for indoor as well as outdoor settings. Indoors, you may need to think about artificial light sources as well, including candle light, modern electric light and all the different kinds of light found in different settings. If there is a fire in a hearth, it is the job of the lighting designer to find a way of presenting this effectively, perhaps including naturalistic flickering of flames. Here there is a secondary light source – the fire – and this would need to be included in the lighting plan.

Devised

A lighting designer for devised work not only has to consider the same points as for scripted plays, but often has the opportunity to develop very creative approaches. Strong side or overhead lighting may be effective for dance or abstract themes, and projected images using a laptop and projector can have a strong impact. Lighting can change the mood of a piece from moment to moment as well as help keep up the pace of the work if it is operated efficiently.

Figure 4.10 Lighting can create surprising effects

Useful tip

Hand torches held under actors' faces might give a suitably dramatic and scary effect if that is all you are able to find.

Figure 4.11 If there are any projections being used, you will need to check your lighting works with them

Sometimes you may wish to use special effects in your production. Smoke machines are now relatively inexpensive and may be hired in some parts of the world. Use them to create mists that can be lit very effectively to create mood and atmosphere according to your choice of **colour filter**, but remember that they can be noisy to use, so think about having music to help disguise the sound. Flames can be suggested by using an **effects projector**. Explosions can be created using pyrotechnics and the use of snow machines can produce a gasp of surprise from an audience, for example, when snow appears in the musical *A Christmas Carol* to create the seasonal atmosphere of Dickensian London. The important thing is to make sure that the effect is appropriate and that it fits the agreed directorial concept.

Key terms

Colour filter: a coloured film placed in front of a lantern to change the colour of the light on stage (also known as a gel).

Effects projector: a device used to project an image from a rotating glass disc to give the effect of, for example, clouds, flames or rain.

Design and technical work

Practical lighting

This section gives some hints on how you might work if you have access to a suitable lighting **rig** and have an opportunity to rig **lanterns** and operate a **lighting board**.

These skills will help build a strong understanding of how stage lighting can enhance a performance. The practical experience gained as you **focus** and angle lanterns or experiment with different **gels** will help inform your answers in the written examination, as you will know exactly what works and, perhaps more importantly, why and how.

> **Key terms**
>
> **Rig:** hang the lanterns in the correct positions.
>
> **Lanterns:** lights used to illuminate a set.
>
> **Lighting board:** a control desk for lighting.
>
> **Focus:** concentrate the lights onto a specific area of the set.
>
> **Gel:** a coloured film placed in front of a lantern to change the colour of the light on stage (also known as a colour filter).

Figure 4.12 *Fitting a gel*

> **Activity**
>
> 1 If you are working with a set designer, you can do this activity at the same time as the Section 4.7 'Set design' activity on page 86. If not, try this activity yourself, using a cardboard box model of the set.
>
>   - Cut holes in the top and sides of the box model to shine light through.
>   - Try out different points of **focus** on the set.
>   - Try different **gels** to change the colour and mood on your model set.
>
> 2 Keep notes of all your ideas and make detailed sketches.
>
> 3 Check that all equipment available to you is in good working order.
>
> 4 Research companies which will sell and hire equipment and special effects. What is available? How might you use it?

As the work develops

- You will need to watch and make notes on the development of the characters.
- Keep notes of the location(s) being suggested.
- Discuss your ideas with the group and suggest how the lighting will help.
- Get to know your equipment and make sure it is all in working order.
- If the actors are presenting short pieces, light them.
- Experiment with lighting actors' faces without shadows.

During rehearsals

Decide on the stage form (see Section 4.7) and what the set will look like, and begin to set it out in your drama space. Keep really detailed notes and diagrams. It is a good idea to tape the position of your set on the floor of your drama space as you do not want to have to re-rig your lanterns if, for example, furniture is in the wrong place. You will need a very detailed scenario as there will not be an actual script for this work.

Experiment with the effects you can create. Try out different coloured gels and note the effect on set and costume. Use **gobos** to see if you can create different locations and try focusing spotlights or using **barn doors** (folding flaps fitted at the front of a lantern to limit the spread of light) to suggest light falling through a door or window, as in the scene from *The Crucible* shown at the start of this section (Figure 4.9).

You must keep records of everything you do so that you can write an accurate **cue sheet**, which should have numbered cues, channels, levels and timings. It should be clear enough for someone else to operate it if necessary.

You should be completely ready for the technical rehearsal where you will have the chance to adjust and develop your design and the timing of each cue.

Show:			Date:		Op:
Page no.	Cue no.	In time	Out time	Description of cue	Action on stage

Figure 4.13 An example of a lighting cue sheet

> **Did you know?**
> To create a prison scene, use the barn doors to project an image of light from a window falling on the stage floor, and a gobo with vertical lines will give the effect of prison bars.

> **Useful tip**
> If other groups are working in the same space, use a different colour of tape for each group when you are marking out your set.

> **Key terms**
> **Gobo:** a small, perforated metal sheet placed between the lamp and the lens of a spotlight to project an image on to the stage.
> **Barn doors:** generally four flaps (two opposing pairs) on the front of a lamp that can be opened or closed to adjust the width or shape of the beam of light.
> **Cue sheet:** a list of the lighting changes throughout the production.

4.3 Make-up

Make-up is the painting of an actor's face to give realistic detail of the face when seen under stage lighting or to create a special character effect. Make-up also includes any changes to the hair of the performer that may be necessary.

There are different types of make-up you could experiment with:

- **Straight make-up:** this focuses on wearing foundation, shaping, highlighting and emphasising the facial forms or eyebrows, eyes, nose and mouth. The character you are portraying will determine how this is applied depending on their age, health and complexion.

- **Character make-up:** this focuses on adding to the actor's features e.g. adding facial hair, changing the shape of the actor's nose, or adding warts or moles. You characters may also need a bruise or scar.

- **Fantasy make-up:** this could include non-human characters such as animals or mythical beings.

Research different materials and products

You will need to find out about the different kinds of stage make-up that are in use, even if you are very familiar with one particular kind. One health and safety factor is the fact that some people are allergic to certain products such as greasepaint, so it is important to be able to offer alternatives. If you are new to this skill you must also find out about basic make-up hygiene.

Figure 4.14 Kate from Rhubarb Theatre's Cooking with Kate *made-up as an older person*

> **Remember**
> Make-up is not just for the face – neck, arms and hands may also need make-up to complete the character.

> **Link**
> www.stagemakeuponline.com

> **Remember**
> Make-up can help make your performances more convincing, especially if you are playing an older character.

There are three main types of make-up which are available for you to use:

- Greasepaint: this is found in sticks and tubes and is the most commonly used stage make-up in professional theatre.
- Crème make-up: this is found in tubes and cakes and is lighter and smoother to wear. It is also less greasy.
- Water-based make-up: this is applied with a damp sponge and is good for covering large areas of the skin in one colour.

Your make-up kit will also need pencils, liners, powder, hair, putty, cotton wool and cleaners to remove any initial grime from the skin. Some performers also use false eyelashes when working on a stage with a large auditorium. Ensure you also have a proper make-up remover that is able to cut through thick make-up.

During rehearsals

You will need to be flexible, as the structuring of the work may mean that you will have to create a make-up change during the performance. Make sure that your kit is in good order and make sure that you have a backstage area to do changes.

You will also need to be responsible for any body make-up. You will be able to enhance your designs with the use of wigs and/or hairstyles, so make sure that you discuss your ideas throughout the rehearsal period with the performers and other designers.

At the technical rehearsal, make sure you can rehearse any make-up changes. Don't forget to clean the make-up off properly, and allow time for this.

At the dress rehearsal, run everything as though it is an actual performance.

Activity
Conduct some research into what should form your basic make-up kit.

Useful tip
Always try to have rehearsal time to try out your ideas. You can do lots of sketches and diagrams, which will help you explain your ideas and will provide evidence of how your ideas developed.

Figure 4.15 Make-up can add a striking effect to your production

You should experiment with a range of different types of make-up from greasepaint, water-based make-up and crème-based make-up. You also need to consider how the make-up can be successfully removed from your actors. You must always thoroughly clean all materials and brushes at the end of the make-up session.

Design and technical work

Make-up design for *The Crucible*

Your first make-up design could be for Abigail Williams. She is described by Miller as being "strikingly beautiful" and that will include her skin. A naturalistic make-up is required here to emphasise her natural beauty. Make-up of any kind would be forbidden in Puritan society.

In contrast to Abigail, to show off your make-up skills appropriately, your next make-up could be for John Proctor at the end of the play after he has been tortured and has spent some considerable weeks in prison. Experiment with using make-up to create the appearance of wounds and bruises.

Figure 4.16 *Abigail's naturalistic make-up*

> **Remember**
> Before you begin to apply make-up, it is important to patch test the make-up you are using on the actors to ensure there is no allergic reaction.

Figure 4.17 *John after his imprisonment*

Designing make-up for a devised work

Actors often create naturalistic make-up on themselves, so it is a good idea to give yourself time to experiment with your own character's make-up. The make-up that some people wear every day does not have enough depth of colour when worn under stage lights and often at a distance, so even naturalistic effects will look unnatural when seen up close.

As the work develops

- Keep notes on all the characters as they develop.
- Look out for any opportunities for the use of make-up.
- Get your make-up kit in order and experiment with different types of make-up.
- Research the types of make-up you will need.
- Find time to practise on a couple of people in the group and on yourself.

During rehearsals

Choose the people you will want to make up. As you need contrasting characters, choose an older character and perhaps you'll be lucky and have someone who needs a scar, a wound or a beard.

Decide whether you are going to use greasepaint, in which case you will need sticks for foundation, and liners. You will also need powder to set the make-up, and removal cream. If you use cake foundation you will need to apply it with your fingertips or a sponge, and if you use a cream foundation you will also need powder.

If you are working in a small performance space you will need to be subtle in your application. Check your make-up under the appropriate lighting. When you are ageing a character, think about how their skin tone will change, and look for the natural places on their face to see where their wrinkles will develop in time. Remember that wrinkles are not lines but little furrows, so will need to be shaded and highlighted to create the effect. A product called Wrinkle Stipple is a light latex, which may be useful.

If you are making a beard or moustache, choose crepe hair of the appropriate colour and be sure to use spirit gum and the appropriate remover. You should be completely ready for the technical rehearsal.

> **Did you know?**
> Traditional stage make-up was called 'greasepaint'. It was a mixture of zinc white, yellow ochre and lard!

> **Useful tip**
> Health and safety are very important. Be sure that everything is perfectly clean before and after use. Make sure you are aware of any allergies an actor may have. It is best not to share make-up products, brushes or sponges with other people as there is a risk of spreading infection.

Figure 4.18 Make-up applied pre-performance

Work as part of the group

During the group planning meetings and rehearsals, your knowledge of make-up should enable you to contribute ideas for presentation, and you will need to keep a record with notes, sketches and photographs of how your make-up may change and develop. All theatre involves **illusion**, and make-up can play a vital part in creating and sustaining this. A particular challenge in school productions of any sort is the fact that the actors often have to play characters much older than themselves. Effective make-up can make actors seem more convincing and can help them get more quickly into character. Changes to an actor's appearance might have to be extreme, using **prosthetics** to change the shape of a nose, for example. Even if your production does not require this, it will be a useful part of your research notes if you can demonstrate your knowledge and understanding.

Some things in your make-up kit you will use many times, but you must always wash your brushes thoroughly and allow them to dry before putting them away. This is to prevent any skin problems when reusing your tools.

Another consideration as a make-up designer is where your audience is sitting in relation to the actors. For instance, if the audience is in the round, they are positioned close to the performers; if they are on a more traditional apron stage with a proscenium arch, the audience will be far away from the performers (see Section 4.7), so the make-up will need to be stronger.

Key terms

Illusion: anything that deceives the senses by appearing to be something which it is not.

Prosthetics: artificial body parts; an example in stage make-up could be nose-putty, moulded to change the shape of a nose, then coloured by make-up.

Useful tip

Make sure that your make-up fits the style of the production.

Figure 4.19 *Make-up must be visible from a distance*

Consider also the type of work you are producing, for example, in physical theatre the actors will be moving around the stage and physical activity will cause them to get hot, particularly under the stage lights. This has to be carefully considered by the make-up designer. The physical aspect of the production may have a direct effect on the type of make-up used, and this will have to be carefully explored.

> ### Activity
> 1 Experiment by using different types and colours of make-up on different skin tones.
> 2 To explore the differences in make-up styles and techniques, try making up half of your face (or another actor's face) in a fantasy design, and the other half in a naturalistic style.

> ### Useful tip
> Watch some online tutorials, for example on YouTube, to help you learn basic stage make-up skills, and how to use different products, tools and techniques.

The designer must practise the application of the make-up on the actor and then allow them to use it during the rehearsal process. This practice will come through experimenting with the different types of make-up you have researched.

On the written examination, you may be expected to discuss how your use of make-up contributed to the effectiveness of your performance piece.

Figure 4.20 An example of the hard work and effort that goes into fantasy make-up

Design and technical work

4.4 Masks in performance

Masks have been used effectively in theatre for thousands of years. Why not think about using them to enhance your practical work? They may even solve some problems, such as a shortage of actors, representing animals, or fictitious or mythical beings.

Doubling

Masks are an excellent solution to the problem of a small cast and a large number of characters. This is especially true of non-naturalistic dramas such as *The Caucasian Chalk Circle* or *Animal Farm*. They can be used to suggest animals, robots or fantasy creatures and can be passed from actor to actor during a performance to give flexibility.

Visibility

Masks are made to be seen. Make sure that your design is clear, unfussy and can be "read" at a distance. A simple design is often the most effective, especially when you bring it to life through your movement and gesture.

Audibility

In your production you need to ask, "Can I still be heard when wearing the mask?" You may need to reshape the mouth, or indeed consider removing the whole of the lower jaw in order to make it possible for you to be heard. To save yourself problems, do try the mask on at several stages in its construction – never spend hours making something, painting it and then find that there is no way for the actor to speak, breathe or even see. Masks must be practical, first and foremost.

Figure 4.21 Actors wearing masks in a production of Animal Farm *by George Orwell*

> **Remember**
>
> The mask is the actor's tool, not the other way round. As an actor, you should not rely on a mask alone to create the character, but instead use it as an aid in your performance to help create a believable character.

Characterisation

The face depicted on a mask can offer an instant indication to an audience of character. The actors in Ancient Greece used to study their masks before a performance to give themselves an insight into the character to be played, and this is still a good idea for modern performers.

Figure 4.22 Greek actors prepare for a satyr play

Did you know?
Ancient Greek theatre masks had mouth pieces that were designed to amplify the actors' voices.

Figure 4.23 Masks used in an interesting way

Performance style

When you wear a mask, you can no longer use your facial expressions to convey emotions. Instead, you will have to use your whole body. Gestures need to be clearer and larger and posture will have to be more expressive. The Activity box below will give you an opportunity to experiment with this for yourself. Good luck and enjoy your experiments – you are now part of an ancient theatrical tradition!

Activity
Materials

- Access to the internet
- Access to reference books
- Sheet of thin card, size A3
- Paints
- Scissors
- String.

1. Research Greek theatre masks and find images of a tragic mask and a comic mask. Look at the images carefully and list the differences: shape of the mouth, eyes, distinguishing features, and think about what it is that makes them have character.

2. Design a tragic and a comic mask for a modern performance.

3. Take a sheet of thin card and cut out two ovals the size of your face and mark the position of eyes, nose and mouth.

4. Draw your finished designs onto each of the ovals and colour them if you wish. Cut out the eye-holes (and nose and mouth spaces as well if it will make it more comfortable). Tie a short piece of string to each side – a stapler will help do this quickly. You now have two masks with which to experiment.

5. Try wearing the mask and watching yourself in the mirror or, better still, work in groups of four with two wearers and two watchers to give advice and direct the following scene.

The scene

A sad person is standing in the street when a happy person comes along…

- How do they greet each other?
- Are they strangers?
- Does the happy person want to help?
- How is this to be shown by the actor?

Remember that the mask only registers when facing the audience – what advice should the watchers be giving?

The scene will be short but you will have had an introduction to using masks.

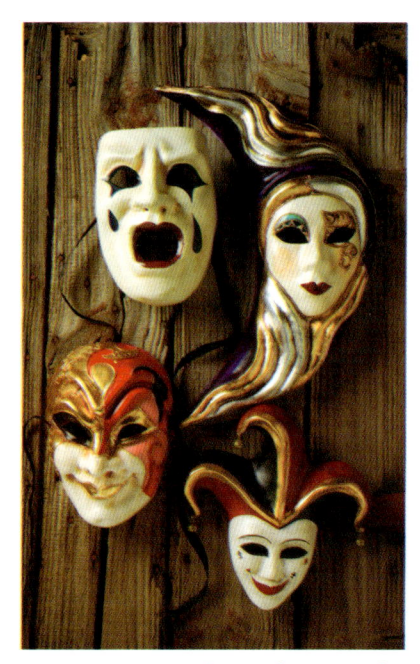

Figure 4.24 *Traditional Commedia dell'Arte, Venetian-style masks*

As your work develops

Consider the following questions:

- Why are we using masks in this production?
- What is the directorial concept?
- What is this mask meant to show?
- What is the style, period and genre of the piece?
- Who is to wear the mask?
- How comfortable will it be?
- What does the actor have to do in the production?
- How will it be lit?
- What costume will be worn with it?
- When can you have fittings and rehearsals?
- When is it to be ready?
- How many times will it be worn?
- How durable will it be?

Consider other periods and styles of mask such as:

- Commedia dell'Arte
- Noh, kabuki, bunraku
- Masks used by modern companies such as the Gateway Mask and Mime Theatre in Atlanta, USA.

> **Activity**
>
> Research different types of mask and techniques for mask-making, such as plaster, papier-mâché, and Venetian masks.

Figure 4.25 Japanese kabuki masks

Design and technical work

4.5 Properties

Stage properties (props) are any objects, apart from the set, that are used by the actors in a production. They are important tools for the actor and this section will help you to manage their use. Carefully chosen props can enhance a production and give confidence to an actor.

You may find a reference to properties in questions on your written examination, so always consider:

- What item of property is most appropriate for the scene you are working on?
- What is its size, shape, weight, material, style, and historical period?
- How is it to be used?
- Why is it specified by the writer or the director?
- What effect or impact will it have in the performance?

These are the activities you will need to be able to carry out when you start work on a production.

Activities

- Choose a play text and make a note of every prop that is mentioned in the text. In a devised piece, for each prop, you would make notes in a rehearsal and ask the group what period, style, shape or size it should be and how it is to be used.
- List all props carefully on a sheet of paper with page or line references next to them.
- Write 'Personal' and the character's name against those that are used solely by one character. The actor will have to be responsible for looking after these and may even be able to find them for you.
- Write down possible sources of the props you think you will be able to find.
- Write 'Make' against those props you will have to make yourself.

Research widely

You will need to show that you know how and why props are used in the theatre and know how to obtain and make them. As productions can cover a range of historical periods you will need to know about period style so that your items fit in with the concept and are convincing.

You must be practical and organised and take an active part in the planning. Take notes and make suggestions, research the chosen topic and the period in which your production is set. At rehearsals keep notes on every object that is mentioned and list who uses them and when. This will be the basis of your props list.

Personal props

Personal props are props used by a particular actor to help create the character, and should be set out in the dressing room before a performance. Some of these will be provided by the actor, others you will have to find. Get into the habit of making them or a suitable substitute available from the start of rehearsals, and of collecting them in and recording them on your props list.

Making props

Some properties you will have to make yourself – these are called **makes**. Learn methods of construction such as using chicken wire and papier-mâché or expanded polystyrene. Your technology and art departments should be able to offer you help and advice. Before you start, make sure you have technical support such as materials, tools, work space and advice.

Props can be borrowed or hired. Friends, family or local businesses can sometimes be willing to help for a mention in your programme. Take a receipt book and record full details of the lender, the value and condition, and give a receipt. State when you'll return it, and do so promptly.

Finding props

The prop you construct has to look realistic on stage. It will be used in rehearsals and must be sturdy enough to last through the rehearsals and for the run of the show without falling apart or looking tatty.

Figure 4.26 A variety of props

> **Remember**
> Always make sure that your make-up or props fit the style and needs of the production.

> **Key term**
> **Makes:** items that cannot be found or borrowed and must therefore be made.

> **Activity**
> - List the props that you have decided on so far, with a note of where they come from, who uses them in the production and which scenes they are used in.
> - Sketch the items you will make, and draw up a list of materials you need.
> - Get feedback from the performers on your ideas.

4.6 Puppets

Puppetry is a medium that has a rich history covering many cultures. So why not consider using puppets in your practical work?

Types of puppet

There are many different styles and forms of puppet; these are a few of the main types:

- Finger puppet – probably too small to be seen clearly on stage
- Glove puppet – also called a sock puppet or arm puppet
- String puppet – also called a marionette
- Black light puppet – uses ultra-violet light with a back-lit screen
- Shadow puppet – an ancient style used in many cultures
- Rod puppet – worked by rods of wood or metal.

Some forms rely on the puppeteers being hidden from view; others work just as well if the audience can see how they are manipulated. The important thing is for the audience to accept them as characters with feelings and emotions. Michael Morpurgo's play *War Horse* at The National Theatre in London has been made into a successful feature film. In the live production, the puppeteers appeared on stage but worked so skilfully that the audience focused its attention entirely on the puppets.

Figure 4.27 Puppetry in War Horse *at the National Theatre in London, UK*

Activity

The simplest form of puppet is the sock puppet, a basic kind of glove puppet.

Materials

- One long sock (clean!)
- Two small sticky-back circles of coloured paper, torn from a sticky note, or a couple of the disks used to reinforce the holes in file paper.

Method

Slip a hand into the sock so that your fingers are touching the end. Hold the back of your hand uppermost with your thumb underneath to form a lower jaw. You may need to tuck some of the fabric in towards the palm of your hand to make the mouth. Then apply the coloured dots or paper as eyes and you will have a basic puppet. You are now ready to experiment.

Scenario

- The puppet is a snake (think about the voice and movement that will give this illusion) and you are a person that the snake is persuading to do something that you know you should not do.
- Your first line is "I can't do THAT!"
- Make the snake as persuasive as possible, not only in what you make it say but also by its actions, for example, nuzzling up against your cheek or pulling sharply back from you and appearing to stare at you.
- Remember that part of the illusion of it being 'alive' will come from you and the way you interact with it. Eye contact always helps.

You can also try experimenting with other forms of puppetry, such as shadow puppets in the following activity.

Once you have an idea of how shadow puppets work, experiment with making them larger and with moving body parts. Try different ways of making the joints work so that the animation becomes smoother. Think of ways in which they could be used in a production, for example, in *Macbeth*. Experiment using an overhead projector to show images.

Keep notes of your experiments and your progress. Comment on your directorial concept, what you were aiming to achieve, what you did and how successful you were. These notes will be useful for revising your written paper answers, and you may need to refer to them as you experiment further.

Activity

Start simply with a sheet of white paper or fabric stretched over a frame with a reading lamp shining on it.

- Cut out your puppets from pieces of card.
- Attach thin sticks to the puppets with adhesive tape.
- Shine a bright light on the side, away from the audience, and you will be ready to place the images on the screen.
- Make sure that your puppets are placed flat on the screen, otherwise they may not be clearly visible.
- Be careful not to stand in front of the light or you will be the shadow instead.

Figure 4.28 Shahrazad brings to life her story of the Talking Bird and the Singing Tree in the Royal Shakespeare Company production of Arabian Nights

4.7 Set design

The design of the set for your work is very important because it fulfils many functions.

The design should help the audience to understand the period of the piece, the mood, the location and the genre. The audience must also be able to see what is going on.

It should also help the actor and the director to portray characters and their relationships on stage by giving them opportunity to use the space, entrances and exits and levels.

When you are devising work, you will be able to develop ideas as you go along, but when you are working on a script, as well as being imaginative you will need to take into account the needs of the playwright.

The basic components of set design are **flats** and **rostra.** Experiment with the ideas given in the Activity box.

This will give you some sense of the moods you can create and, if you find an arrangement that seems to be both suggestive of the mood or theme and practical for actors and audience, it could be the basis of your design.

> **Key terms**
>
> **Flat:** a light wooden frame covered in scenic canvas, plywood or hardboard, which can be painted to suit your work.
>
> **Rostrum (plural rostra):** a portable platform which you can use to create interesting levels.

Figure 4.29 Set design incorporating a red brick wall

Activity

Try the following experiment to see how rostra and flats can work together.

- Find a cardboard box and paint it black inside (or cover the inside of it with black paper).
- Cut out a series of rectangles and paint them white. Add a tab at the back so that they are free-standing.
- Collect or make a series of small boxes and paint them white.
- Now arrange them in the box. You will have a huge number of possibilities.
- Put in a cut-out figure to get some sense of scale.
- Try lighting your arrangements either with lanterns or with torches.
- Try different combinations of coloured gels and light from different directions.
- Take photographs of the different combinations to discuss with your group.

The place where you perform your play will influence a great number of things. The performance space must be suitable for the play to be performed, the actors to act and the audience to see, hear and appreciate the play.

Using your space

- Make a scale **ground plan** of your performance space. Use the scale 1:25 and mark on it all permanent features, such as doors and the lighting rig.
- Sketch on the plan how different types of stage would work and where the audience would be. Turn to the next page for different types of performance space.
- Begin to collect and research sources for the actual set.
- Begin to sketch ideas with samples of colours, materials and textures.
- Create a mood board to keep a record of your ideas and decide what will work best.

You will need to get to know the possibilities of your own performance space. Some schools and colleges have an assembly hall with a **proscenium** arch stage.

This is a stage form which divides the audience from the action like a picture frame and is used in many traditional theatres.

> **Key terms**
>
> **Ground plan:** a scale outline of the set drawn as if from above with indications of flats and furniture marked on it.
>
> **Proscenium:** an arch that separates the stage from the auditorium, common in traditional theatres.

> **Useful tip**
>
> Make lots of copies of your ground plan.

Figure 4.30 Proscenium arch stage, Roosevelt University, Chicago, US

Figure 4.31 Theatre in the round, Hong Kong Cultural Centre, Kowloon, Hong Kong

Many theatres have adaptable or open stages. However, there are a number of different types of performance spaces for you to explore and consider in your design.

Design and technical work

Examples of open stages

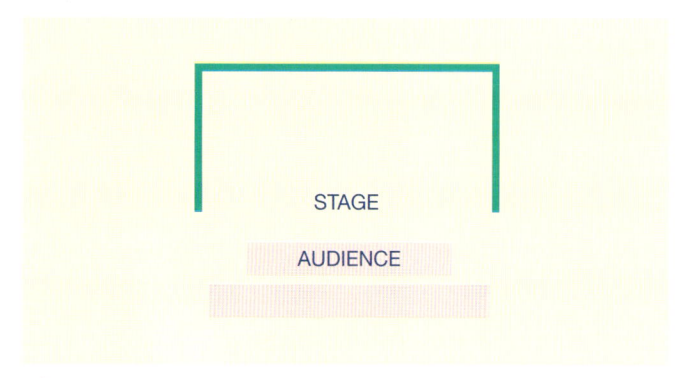

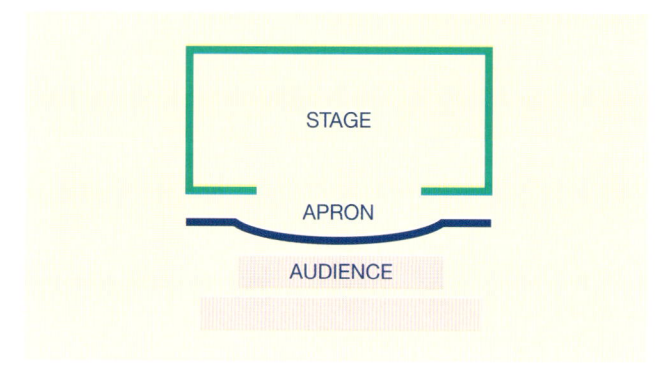

Figure 4.32 *End-on stage*

Figure 4.33 *Apron stage*

- **End-on staging** has the audience on one side. A teacher in front of a class is a simple example of this stage form.
- **An apron stage** extends the performance space beyond the proscenium arch, bringing the actors closer to the audience.
- **A thrust stage** has the audience on three sides and is connected to the back-stage area up-stage. The actors can exit and enter here and also through the audience.
- **Theatre-in-the-round** stages can be any shape but the audience surrounds the action on all sides. The actors enter and exit through the audience.

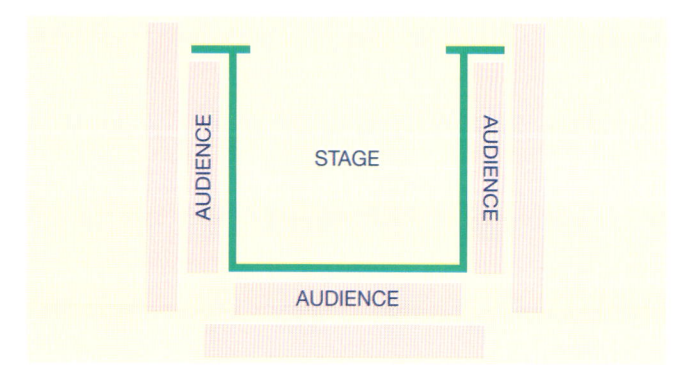

Figure 4.34 *Thrust stage*

Figure 4.35 *Theatre in the round*

Working on a script

When drawing ground plans and sketches, it is important to pay particular attention to the detail that the playwright gives. Consider:

- Where are the entrances and exits?
- Where will your actors stand while they are waiting to come on stage?
- What view will the audience have of the actors as they perform?
- What materials will you use to construct your set design?
- How will your design be created?

You will also need to consider costing, textures, colours and fabrics in your design, and how your design will look when other design elements are considered.

> **Remember**
> Your actors will need space to move around on stage!

Creating your own set design sheet

Activity

By yourself:

- On an A3 sheet, write the title of your scripted production. In the centre, draw a sketch of the set – it doesn't have to be a masterpiece. Add the main colours, using written labels if easiest.
- Around the sides of the sheet, write down notes about the following:
  - General impact and first impressions of the set
  - Description of the set's features, in clear and logical manner
  - Period/Location: Where is it set? When? How does the designer show this?
  - Structure: What form does it take? Why?
  - Levels: What levels are used? How? Why?
  - What illusion is the designer trying to create?
  - What materials do you think the designer chose to create that illusion?
- Each member of the group should annotate their own drawing, referring to as many bullet points as possible. (You could maybe make photocopies for each student and one extra as a main copy for the whole class.)
- Colour-code your comments – use separate colours to show What? How? and Why?

As a group:

- Ask the other members of your group to describe the set as if taking a mental walk around it. Remember to follow the same pattern, for example, from downstage left in a curve to downstage right. Use the correct stage positioning terms so that anyone can follow your description. If you are unsure, turn to Figure 6.2 in Section 6.1.

This exercise will help everyone get into the habit of organising their thoughts and ideas by giving a simple structure. Get into the habit of doing this by applying it to every set you use so that you can memorise them more easily.

The following design in Figure 4.36 takes into account the needs of the playwright, including exits and entrances, use of space and period furniture, and shows the differences in status of the characters.

Period: 1914–1918
Shown by: Oil-lamps and candles for lighting, props – e.g. steel helmets, mess tins costumes – authentic uniforms.

Style: Naturalistic – it made you feel as if you were looking into a real trench dug-out. The accurate detail of costumes and the use of paraffin lamps and candles helped create an enclosed and cramped atmosphere. Ammunition boxes for seats and rough camp beds made it look like a place for working rather than comfort.

Door to the kitchen area

Entrance from surface

Location: A dug-out in the front-line trenches during the First World War.

Shown by:

levels – steps leading up to surface

textures – rough timber, corrugated iron sheeting, timbers to support the roof

colours – earth coloured browns, muddy, drab

furniture – camp beds, folding chair for C.O. only, boxes for the others.

Genre: Anti-war drama, tragedy

Figure 4.36 *Set design sheet for the play* Journey's End *by R. C. Sherriff*

Working on set design for a devised piece

Take careful notes of the rehearsals so that you can be clear about the mood of the play and the location that the set needs to communicate to the audience.

It is likely that in devised work a naturalistic piece will be created, so you may need to collect furniture which gives a sense of how at least one of the characters lives.

Present your ideas to the group and remember to consult with any other members of the group who are also working on a design skill. Try to accommodate the actors' needs in your design but also try to avoid a series of short scenes, and suggest solutions to keep the play flowing, such as a **composite set**.

Any furniture you include in your set design must be suited to the character whose home you are showing and to the period in which the play is set. If you are using flats to suggest walls, they must be painted or papered appropriately, and make sure that you discuss your colour scheme with the actor and with the person responsible for costumes. There will be some items which are needed because of the plot but be sure to dress your set appropriately too.

Health and safety is very important for the set designer. So consider:

- fire precautions, fireproofing and fire exits
- backstage safety for the actors
- front-of-house safety for your audiences.

The photograph opposite shows how detailed the designer's model should be. The play is set in the 1970s and fully reflects the period and the taste of Beverly whose home it is. Note the leather furniture, the ornaments and pictures which **dress the set**, and the wallpaper and colour scheme.

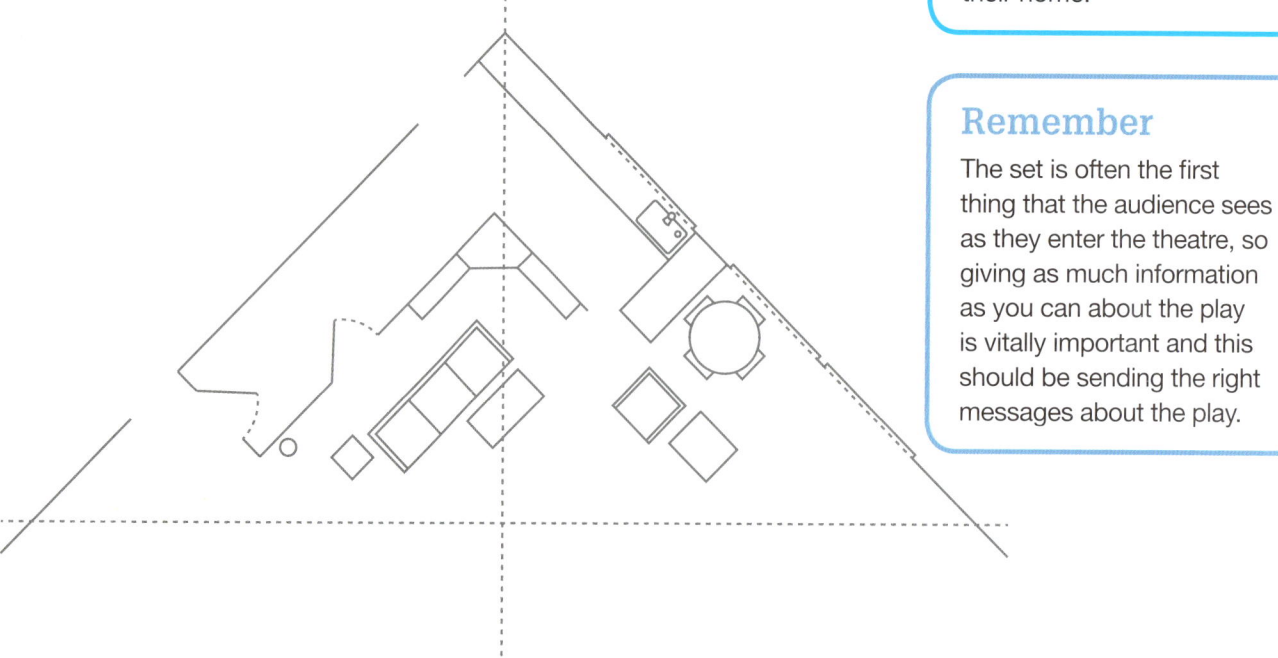

Figure 4.37 Ground plan for set design opposite (not to scale)

> **Key terms**
>
> **Composite set:** a set on which it is possible to perform scenes in different locations or rooms in the same building without having to change the set all the time. *Blue Remembered Hills,* for example, needs a composite set to keep the action moving.
>
> **Set dressing:** including items on the set that add to the overall effect. For example, Figure 4.38 has cushions, plants and pictures.

> **Useful tip**
>
> It would be very useful if, during the hot-seating sessions, you could ask questions that will encourage the actors to think about their environment and how they think they might furnish their home.

> **Remember**
>
> The set is often the first thing that the audience sees as they enter the theatre, so giving as much information as you can about the play is vitally important and this should be sending the right messages about the play.

By now you will have finalised your basic design and may find it a useful experience to make a scale model.

If you choose to make a model, use the scale 1:25. Make your ground plan of the set, taking into account the form of stage you are using (e.g. thrust, end on) and pay attention to **sight lines**. The simplest way of doing this is to mock up your set and audience seating, get people to walk the set and others to sit at the extreme points of the seating. When you are satisfied that the set works, put your sight lines onto your ground plan. Then make sure that the actors are aware of this. When you tape the set out on the floor of your space, tape in the sight lines too.

Mark a **setting line** by a snap line – a chalked string pulled taut between two people at floor level and twanged on the floor to leave a neat chalk line. Mark the centre line – fold the string in two to find the mid-point. Make a snap line from this point at 90 degrees to your setting line. You will now have two axes corresponding to those on your plan. Transfer measurements from the plan, multiplying by 25 to scale up. If you find measuring difficult, perhaps you could ask your maths teacher for practical advice in applied geometry.

Be aware that there might be a need for projections at some point, so experiment with different types of screen and different directions of projecting. You will be able to advise the group of the best way of getting the points across and be able to use the screen as part of your design. If you are working with a lighting designer, you must consult with him or her.

> **Key terms**
>
> **Sight line:** what your audience can see on the stage; sit at the extreme ends of the front row to work it out.
>
> **Setting line:** an imaginary line marking the furthest point downstage that scenery may be set, usually just above the tabs (curtain) line. Marked on stage plans it offers a reference point for placing flats, scenery and other items on stage.

> **Useful tip**
>
> Take photos of your work under the lighting.

Period: 1970s. Shown by colour scheme – brown and cream, leather furniture

Location: Home of Beverly.

Style: Naturalistic. Interior – warm and comfortable. You feel as though you are looking into someone's house.

Notice that there is plenty of down-stage space

Genre: Black comedy

Figure 4.38 *Designer's model of the set for* Abigail's Party *by Mike Leigh, directed by David Gridley*

Design and technical work

Be practical

Your designs will have to be practical. Show that you have thought about the needs of the production. This means that you will need to think about how you would build the set, if you had the time and materials to do so. The set will have to be used by actors in performance, so it needs to be able to help them, giving them clear entrances and exits as well as space in which to move.

Consider the following:

- how can you use flats and rostra to create your set?
- is your set fixed, or do you need to be able to transport it easily?
- what stage form will you have?
- how many different scenes do you need to create?
- what furniture/props will you need to set the period and location?
- what movement will your actors need? Where are the entrances and exits?

Fit in with other design elements

Your designs for the set will have to fit in with the lighting, costumes, puppets, masks and properties. As a group you will have discussed style and made decisions. From that point, all the design team will need to work together to make sure that style, materials and colours fit together and do not clash with each other. The production may be set in a particular period or show different moods or atmospheres and you will have to research ways of showing these through your designs.

Look at this photograph of *The Crucible* in performance (Figure 4.40).

Figure 4.40 The Crucible, *Royal Shakespeare Company production, Stratford*

Note how the lighting enhances both the upstage wall of the set and the grouping of the actors, and shows the sombre and serious mood of the scene.

> **Remember**
> If you are designing for devised work, you will need to be flexible because the nature of it means that new ideas will be buzzing around throughout the preparation period.

> **Remember**
> Take into account live productions you have seen in the theatre and be aware of any ideas that you could use and adapt.

> **Useful tip**
> Seeing the work of other people can always help with ideas for your own designs.

Keep to the concept

You will also need to think about the concept of the production, the basic ideas that you want to communicate to the audience, and how the set can help convey this to an audience. You could use distinct images that the audience will immediately recognise, for example, the Eiffel Tower in Paris.

In this photograph of *The End of Everything Ever* in production (Figure 4.41), you can see how a large wooden wardrobe was adapted to form the basis of a set. The concept of packing up and moving on was also suggested by the wardrobe and the material – rough wood also hinted at the rough life ahead. It was used for entrances into interiors and, as can be seen in Figure 4.41, as a train cabin as the child begins her journey away from her family.

> **Remember**
> Be aware of:
> - health and safety issues
> - the needs of the actors
> - the needs of the audience
> - period, mood, atmosphere, style.

Figure 4.41 *In* The End of Everything Ever *by NIE Theatre, the whole design was based on a wooden wardrobe*

4.8 Sound

Sound refers to any extra sounds that are not made directly by the actors. One of the most important aspects of using sound is that your actors must be heard. Your sound effects must also come in on time, be of the right length and volume, and appropriate to the performance. Using appropriate sound effects could enhance your production and make it seem more professional.

As the work develops

Try the activity ideas in the margin during the development phase of your devised or scripted work.

During rehearsals

You will have the chance to make all sorts of ideas work. You may be able to use personal microphones or microphones on stands. You will be able to decide where your speakers should be placed for maximum effect. Some sounds may be effective coming from behind the audience.

Keep the **cue sheet** up to date and keep copies of everything you record. If you are recording performers in your group, make sure you set aside time to do this early on in the rehearsal period so that you will have time to edit and refine it. Keep a clear plan of where everything will be and take note of health and safety issues. Where will your equipment be and where will you operate it from? Work with the lighting designer or operator so that you can time cues together. Work together on the opening sequence to establish the mood from the very beginning.

At the technical rehearsal, finalise your cue sheet and make sure you have details of volume, length of the cue and which speakers you will use. At the dress rehearsal, everything should run as though it is a performance.

You may wish to use sound to create, for example, the effect of being in the countryside. You can do this by playing recordings of birdsong or cicadas. These are known as **diegetic** sounds as they belong to the actual place being evoked. **Non-diegetic** sounds would be music associated with the countryside, or a commentary saying "And now we move to the countryside…"

Figure 4.42 Water *by Filter Theatre Company – the sound equipment does not have to be hidden from the audience*

Activity

- Collect and record songs and music appropriate to the production.
- Note any sound-based moments in productions you have seen which you would like to try out.
- Be aware of the needs of the performance – any dance numbers or fight sequences.
- Try composing and recording music and sounds with others in your group.
- Experiment with putting a soundtrack together.
- Experiment with recording individual voices and choral work. Add a backing track.
- Research sound effects on CDs or online, and try recording your own effects.

Key term

Cue sheet: a list of the sound changes throughout the production (also known as a sound plot).

Activity

Research the theatre company Stomp and how it has used sound in its latest performance.

Link

www.stomp.co.uk

Sound effects add to the production experience as a whole and are sometimes a vital ingredient in the success of the production. They also add to the production by enhancing and creating mood and atmosphere. The sound effects can be pre-recorded or produced 'live'.

Musicians can be used to produce live sound to enhance the quality of the performance. Some theatre companies create the whole of their performance piece from live sound. 'Stomp', for example, are known for using items such as brushes, bin lids and drums.

The choice of music and the sounds integrated into the performance are really important and will be crucial to its success. This should be researched from the outset and form part of your early planning. You and your group should experiment with a variety of music and sounds. Are you using pre-recorded music or abstract sound or a combination of both? Another possible consideration is the creation of a 'soundscape' where abstract sounds and key words combine to produce your sound stimulus. This will help to underpin the narrative which links to the action you perform.

Here are some ideas to consider when planning and rehearsing your work:

- Join in the actors' hot-seating sessions and find out about the musical tastes of the characters.
- Keep notes of the action on stage and anticipate effects you might need.
- Listen to lots of music and record some that might be useful.
- Research sound-effects CDs.
- Experiment by comparing pre-recorded sounds with those you can use live – which are most successful?

Sound design for scripted work

As a sound designer for *The Crucible*, you might prepare and present a **sound plot**. Your work could include a practical demonstration of the sound effects necessary for the production. You would need to produce a sound plot, annotated script, diagrams and considerations of health and safety issues.

The Crucible offers you a wide range of sound effects necessary for the success of the production.

The beginning of Act One gives much scope for experimentation, particularly the section where John Proctor and Abigail Williams are arguing over the "sleeping" Betty. Reverend Parris has gone downstairs to speak to the congregation to lead them in a psalm, and this is heard coming up from below. The words "going up to Jesus" are heard, and Betty takes this as her cue to wake up and start screaming.

The volume of the psalm is a crucial factor in the successful execution of this scene, and great skill will be required here to add to and build on the tension of the scene. This culminates in the entrance of Rebecca Nurse when "everything goes quiet".

As a sound technician, you have to decide which sound effects are live and which are recorded. You also have to research the pros and cons of live and recorded sound effects. The direction of the sound effect is also very important – as the stage directions indicate that the sound comes from below, how might you achieve that?

Useful tip
A solo violin could evoke a poignant moment.

Did you know?
You need to be aware of copyright restrictions on all music and recordings, especially if you are giving a public performance.

Remember
Accurate cuing of sound effects is crucial to the success of the production. If, for example, a doorbell sound effect comes after the actor has pressed the bell, a dramatic moment in a production can become one of humour and destroy the atmosphere of the play.

Key term
Sound plot: a list of sound cues and levels in running order.

Useful tip
It is necessary to carefully plan and rehearse this scene, including the use of the sound effects. If the sound effect is too loud, the dialogue of the scene will be lost.

Recording and editing sound effects:

Sound effects can be added to your performance to enhance the production element of your piece. Access to suitable resources is important to ensure successful outcomes and you will need to show knowledge of: the use of live, directional and recorded sound.

Live sound effects are both fun and easy to produce and if the sound is simple they are also most effective in practical terms. These sound effects should be produced backstage and used in performance in real time if at all possible. It is particularly effective if the sound effect is required for a short period of time; for example a gunshot or a door bell ringing.

Once you have read through your script you can highlight where a sound effect could be added and then experiment with simple resources. You can roll dried peas or beans or rice in a drum to create the sound effect of rain.

Breaking dried sticks, crackling dried paper, sawing through a cabbage can all give amazing results.

You can also move on to record your own sound effects for use in your performance-swish water around in a bucket, record it close up with your microphone and you will have an excellent sound effect of water lapping on the shore. Here you can experiment to get just the sound you want.

Directional sound is important to convince your audience that the sound is coming from the right place on the stage. For example, if a telephone rings then the sound must come from the source of the sound to make the effect believable.

When a sound effect is needed for a long time on stage, it is probably best to record it. With all recorded sound effects, you must make sure that the quality of your recording is as good as possible.

Your recorded sound effect must be convincing so listen to it from where the audience is sitting to ensure its credibility.

You will also need to carefully annotate your cue sheet to ensure accuracy in performance terms. Remember that your sound effect must not drown out any speech on stage, it must be there to enhance the performance not overwhelm the dialogue.

Your sound effect must also be historically correct so research your time period carefully before adding your sound to the performance.

Use commercially produced tapes wisely and make sure you edit your tape carefully to ensure success.

The role of sound designer

Of all the design elements, the sound designer has a really challenging and exciting role in any production. It is possible to link the acts or scenes of your production with a soundscape or sound narrative. You must create a working plot that is detailed and accurate to help you in your task.

As a sound designer, you will need to be aware of the needs of the play, but also to suggest your own ideas about how you can indicate the moods of different scenes or sections. You may be asked for specific sound effects, such as a chiming clock, but you also need to be aware of how else you can use your skills to contribute to the overall effect.

Figure 4.43 The sound technician must communicate with the production team at all times

Show:					Date:		Op:	
Page no.	Cue no.	Device	Disc name	Track no.	Track name	Track length	Track description	Action on stage

Figure 4.44 An example of a sound cue sheet

Managing sound for the performance

A well-produced soundtrack can enhance any production, but there are a number of areas to consider to avoid problems.

Directional sound is very important, as the sound must appear to come from the right place on the stage.

The sound must be there to enhance the action, not overwhelm it. Some points to consider include:

- Think about and suggest music to begin and end the play. Will it be playing as the audience enters the theatre space?
- Try your sounds through speakers both backstage and front-of-house. Make sure they are not too loud or too quiet.
- Microphones and speakers must be in good order for the duration of the show.
- Make sure that you have an up-to-date scenario or script and note where the sound will be needed.
- Your cue sheet should have numbered cues and should show when the cue happens – it may be with the lighting, on an entrance of a character or a line spoken. It should also show how it fades in and out, how many seconds it will last, how loud it is and which speakers will be used.

You should be completely ready by the technical rehearsal, when you will be able to finalise your cues: their placing, length and volume.

Figure 4.45 A sound desk

> **Remember**
> - Always follow health and safety rules.
> - Tape down loose cables.
> - Keep notes of what you have done.

> **Useful tip**
> Always try to fade effects and music in and out, unless it's something sudden like a sharp bang.

> **Useful tip**
> Be ready to discuss how your use of sound contributed to the effectiveness of your performance piece.

> **Activity**
> Research the different types of microphones you can use in a performance. Find out what each type does and which microphone is best for your needs.

Unit 5 Developing your physical skills

Physical theatre is a **genre** or style of performance which makes use of the body as its primary means of performance and communication with the audience. It is a visual form of theatre where the actor concentrates on the use of the body, shape and position, facial expression, movement, gesture, posture and voice.

Objectives

In this unit you will learn to:

- consider the development of physical theatre
- consider what makes physical theatre different from other forms of theatre
- consider how the use of your physical skills enhances the character you are portraying
- use your creativity in producing a response in a physical way
- warm-up and exercise your breath and voice
- use your voice in a safe and effective way for performance.

Key term

Genre: a category of art, music or literature; drama is a literary genre. Drama is further divided into types of production such as tragedy, comedy, farce and melodrama. These genres in turn can be subdivided.

5.1 Using physical skills in performance

Physical theatre focuses on the telling of a story, or narrative, largely through physical means. For example, an actor may represent other things, such as objects of emotions (no matter how simple or complex) using the body. Within your coursework performance, you will need to consider relationships, characterisation, conflict and narrative.

Styles of physical theatre

There are various styles of physical theatre, including physical comedy, mime, contemporary dance, theatrical clowning and theatrical aerobatics.

Some companies only use physical theatre in their performances, but many companies will use aspects of physical theatre as part of their performance outcomes.

> **Useful tip**
>
> Exploring the work of companies who specialise in physical skills will help you in creating your own character in performance.

Figure 5.1 Using body movements to convey comedy

Developing your physical skills

The most famous development of this type of theatre came from the Lecoq School in Paris. Here students followed the method of Jaques Lecoq. His work developed from mask work, Commedia dell'Arte and his interest in physical theatre.

Did you know?
Commedia dell'Arte is a theatrical style developed in Italy from the 16th to the 18th century.

Each actor played a clearly defined character such as an old man or a servant. They were most skilled in improvisation and comedy.

Key term
Stock character: a farcical character in theatre, who reoccurs in a dramatic tradition, and is therefore quickly recognised by the audience and needs no introduction.

Figure 5.2 Harlequin is a stock character in Commedia dell'Arte

In recent years, a most popular company has developed using a pure form of physical theatre. 'Stomp' uses the body, everyday objects – such as bin lids, brushes and spades – and percussion instruments to create a dynamic, exciting and energetic spectacle (see Section 4.8 for more on 'Stomp').

Activities
1 Figure 5.2 shows the stock character Harlequin. Research other stock characters in a theatrical style of your choice.
2 Devise your own scenario based on those characters.

Activity
Collect various household "instruments", e.g. pots and pans and wooden spoons, and create your own simple soundscape to create atmosphere for your physical theatre performance.

Figure 5.3 Physical theatre performed to its fullest by these performers from Stomp

Planning a performance

Before you begin your planning, you and your group must ask yourselves a number of questions:

- Do we have a suitable performance space?
- Where are we going to perform our project?
- What type of performance do we hope to achieve?
- Is the performance space big enough for our needs?

Physical theatre lends itself to producing a large performance on a grand scale. Think of the opening and closing ceremonies of a large-scale international event, such as the Olympic Games. This is physical theatre in the extreme sense with the narrative firmly identified and a target audience of the world!

Figure 5.4 Opening ceremony of the London 2012 Summer Olympics

In physical theatre, the audience must be carefully considered and you should ask yourself: "who is this drama for and what do I want my audience to get from this performance?" The consideration of your audience is crucial to the success of your project. Your performance must have a definite form, with a beginning, a middle and an end, and the development of the piece should be easily identified as you progress.

Developing your physical skills 101

5.2 Using your skills in developing a performance

Your project will obviously be on a smaller scale than some of the performances we have considered here, but a great deal of planning will have to be undertaken for it to be successful. The attention given to the details of the initial planning is time well spent: the better prepared you are initially, the more time you will have later. The **scenario** you and your group develop will help you with the structure of your performance and **stage form**.

If you decide to perform in a large performance space like a sports hall or an arena, you will also have to ensure that you can rehearse in that same space, as a change in location can cause problems when it comes to the performance.

Planning

You will need to plan a rehearsal timetable with the help of your stage manager, to ensure rehearsals are well spent and time is not wasted. Your group will need a reasonable sense of rhythm, creativity and good movement skills. You will also be aware of **dynamics** and how live and recorded sound will be part of your performance planning. You will also need to think about technical and design considerations that will be linked to your performance outcomes.

Your performance

When performing, you need to be aware of the way your body movements can be integrated into your performance piece.

We now consider five basic body movements: stepping, travelling, turning, jumping, and gesture. These movements should be explored through a series of activities and exercises.

Stepping

Think of as many ways as you can of 'stepping'. There is stepping high on your toes, low on your toes, stepping taking wide strides, stepping taking small strides, stepping with your feet flat, stepping using the sides of your feet.

Take notes as you do these exercises to focus on how it feels to move in this way. Observe other members of your group as they try these movements. Which of these (perhaps with some modification) do you think will most suit the character you are playing? Some actors start their characterisation by choosing the shoes for their role and then they can get the movement right.

Travelling

How will you and your group members travel to ensure fluidity of movement in performance?

> **Key terms**
>
> **Scenario:** the summary or outline of the plot of the play and a list of the scenes in order and the characters which appear in them.
>
> **Stage form:** the arrangement of the acting area and audience in your performance space.
>
> **Dynamics:** variations or contrasts in energy, speed and the movement you create.

> **Useful tip**
>
> All work that you produce must be intended for live performance to an audience and staged in an appropriate performance space.

There are many ways we can travel around our performance space and these include rolling, jumping, sliding, or adding shape and form to our travels by, for example, adding cartwheels. Consider how your character or group uses the space in terms of speed of movement and positioning on stage and how this will reflect relationships and conflict.

> **Activity**
>
> Play 'follow my leader':
>
> - Make a line of six people.
> - The first person sets off around the room and walks in their normal way around the room; the five other members of the group observe.
> - After the first circuit, the second person follows and copies exactly the way the person in front is walking but chooses one characteristic and slightly exaggerates it (e.g. the way the person swings their arms or the length of their stride).
> - Each person then joins in turn, exaggerating a different aspect.
> - Once the line is moving comfortably, and all have tried the walk of the leader, the leader joins the back of the line and the next person takes over as leader.
>
> The exercise is repeated until everyone has had a turn as leader. It is an amazing feeling to try to walk in someone else's shoes! This is an excellent and fun warm-up exercise and, although simple in form, it emphasises that the way we walk immediately identifies the type of person we are.

Figure 5.5 The chorus line

Turning

The way we turn and interact with each other defines the movement made. We can pivot or turn while travelling to add shape to the movement sequence. The way we turn towards or away from other characters will also show the relationship and conflict between them. Are you turning towards them expectantly, enthusiastically or angrily? Are you turning away, fearfully, sadly or impatiently?

Jumping

Another way to make movement around the performance space interesting is to add jumps. There are five basic jumps and these include: hopping, leaping onto one foot, leaping onto two feet, leaping from one foot to the other, and leaping from both feet to land on both feet.

Figure 5.6 Chinese Ballet Company, *Swan Lake,* Lowry Theatre Production

Gesture

Gesture is perhaps one of the most important aspects of performance work in physical theatre. Gesture includes functional movements of the body: waves, nods, arm movements, and hand and facial expressions. The use of the face, including the eyes, is particularly important in this type of performance.

We use gesture unconsciously in our everyday lives, from giving directions to emphasising our conversations. We must now use them in an exaggerated form to add to our physical theatre performance. Observe others carefully and consider which gestures would fit your characterisation, show your mood or your feelings about others on stage.

Activity

Make a list of the different gestures you have used and observed in use in your everyday life.

Did you know?

Rudolf Laban was a pioneer of modern dance and movement in Europe. His work even focused on how the body can be taught to save energy when completing physical tasks. His ideas freed actors to find their own rhythms, create their own steps and use their own space.

Activity

1. Read the 'Did you know?' box about Rudolf Laban. Find out more about Rudolf Laban's work and try to find a visual representation of his movement analysis theory.

2. Experiment with Laban's movement analysis theory.
   - As a group, walk around the room in a natural manner.
   - Every 30 seconds or so, take it in turns to call out one of the movement types, i.e. gliding, floating, flicking, wringing, slashing, thrusting, pressing, dabbing.
   - Each time a new word is called out, everyone should continue to move about the room, but use that word to affect they way in which they move.

Physical skills – facial expressions

Some simple exercises are a good introduction. The simplicity of the action and the visual outcomes of the repetition of the movement are where the power of the dramatic form lies.

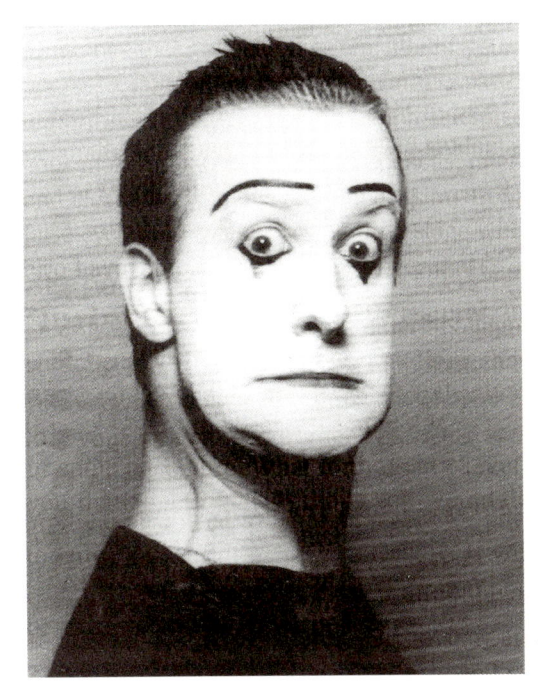

Figure 5.7 Mime artist Rowan Tolley

Activity

Throw a face: a useful game for building confidence in adopting facial expressions.

The group stands in a circle keeping neutral facial expressions. The leader pulls a face, shows it to the group then mimes peeling it off, rolling into a ball and throwing it to a person opposite. The recipient immediately puts both hands on their own face, and on removing them is seen to be reproducing the thrown face. This face is then peeled off and thrown away. A new expression is assumed and this is peeled off and thrown to the next player. Continue round the group making sure everyone is included before the game ends. Other players must try to keep neutral expressions however funny the faces might seem. Make your mime as convincing as you can.

Once you are familiar with the game and the text of a play you are working on you might try a variation in which the leader quotes a line while adopting an appropriate facial expression to go with it. The face is thrown to the next player who receives and adopts it before choosing a new line and face. You can vary the games by having everyone adopt the thrown face, or make anyone who laughs drop out until there is a 'face off' between the remaining two players. Vary the pace to get quick responses or thoughtful ones.

Music and sound can enhance your performance piece by adding structure and atmosphere. You and your group can experiment with a variety of music and simple live sound-effects in your performance.

Extension activity

Divide into pairs. Use a simple mirroring exercise to begin: combing your hair, brushing your teeth, putting on your make-up, or having a shave. Observe your partner and then try to copy the movement in every detail. Do this in slow motion first and then build up to normal speed. See if you can get to the stage where the audience cannot see the changeover point when the leader changes their action.

Remember

Keep your ideas simple. The power is in the repetitive form.

Remember

You can have up to six performers in your group, but it is your individual contribution that is being assessed.

5.3 An introduction to voice work

Voice work is often the skill that is most neglected by students. This unit suggests some activities that will help you to improve the vocal side of your performance.

Warm-up exercises

Breathing and speech are controlled by muscles. Before starting any strenuous activity using your muscles, you first need to warm them up. This is something that professional actors and singers always do before a rehearsal or performance, and it is a good habit to get into at any level.

The muscles you use to make sounds are connected to the other muscles in your body, and tension elsewhere can stop you from using your voice effectively.

Figure 5.8 A theatre group doing vocal warm-up rehearsals

Strain can make the pitch of your voice rise and become squeaky, so it is a good idea to be at ease before you begin. A quick physical warm-up can help you to let go of any tension in your body, as suggested in the Activity box.

Activity

1. Shake your hands from the wrists, then your arms from the elbow and then from the shoulder, as if throwing away the tension.

    Stretch your arms up as high as you can, then drop them to the sides, gently drop your head forward and flop forward from the waist.

    Slowly unwind by rolling up through your spine, vertebra by vertebra, and then repeat the exercise.

2. Roll your right shoulder forward four times, then back four times.

   Then do the same with your left shoulder.

   Finally, do both shoulders together.

   Why four times? Sometimes exercises are carried out to music and 4/4 time is convenient for many tunes. The rhythm is nice and steady, which is exactly how you should be approaching your work.

Breathing

Voice is simply a controlled, vibrating column of air pushed by the **diaphragm** from the lungs through the vocal chords. It produces the human sounds of talking, singing, laughing and crying. If you don't use your diaphragm and lungs effectively, people won't be able to hear you.

Try the following exercise to work on your diaphragm control.

Key term

Diaphragm: a large muscle at the bottom of the rib cage that assists in breathing.

Activity

With one hand on your lower belly, blow out all the air you have in your body. You will feel your belly going in, towards your spine. This is the result of the diaphragm pushing the air out of your lungs. When you have no air remaining, your diaphragm will automatically release (relax). Air will rush back in as you take a breath, because your body needs to take in more oxygen. You will feel your belly expanding again under your hand as you take in this breath. This kind of breathing is automatic and you do it all the time without thinking.

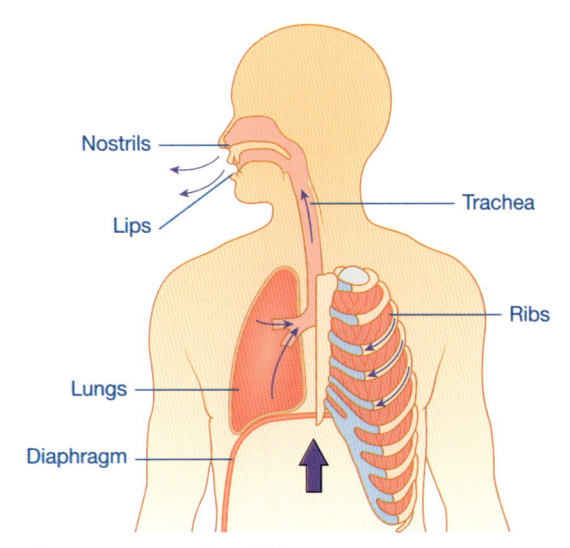

Figure 5.9 *Your breathing apparatus*

If you can harness your breathing properly, you can gain better control over your voice. The following set of breathing exercises should help you to do this.

> ### Activity
> 1. Stand upright with shoulders relaxed and feet shoulder-width apart. Place your hands on your ribs and feel their shape. You will see why they call this the rib cage: it feels like a cage with bars of bone and it protects the lungs inside it.
>
>   - Breathe in (inhale) deeply through your nose and feel your rib cage expand as it moves outwards and upwards. If it doesn't seem to be doing this, try touching your ribs again at the sides and think about filling up from the bottom of your lungs.
>   - Once you have filled your lungs, slowly let the breath out (exhale) in a controlled way (perhaps on a 'sh' sound). When you have used up all the breath, your lungs will automatically fill up again. Now that you are aware of what is happening, you can start to take conscious control of your breathing.
>
> 2. People inhale and exhale automatically. As a performer, you will need to be able to control these actions and, when necessary, be able to take in larger breaths than normal so that you can deliver a long speech without straining. The next stage will help you do this.
>
>   - Standing upright as before, inhale deeply through the nose to a mental count of "1, 2, 3, 4", then exhale to a steady mental count of "1, 2, 3, 4, 5, 6, 7, 8".
>   - Practise this a few times until you are comfortable with it. You can try extending the numbers as you exhale, but keep it steady. Fill your lungs, but make sure that you are not tensing or raising your shoulders as you inhale.
>
> 3. The next stage is to vocalise your breath:
>
>   - Inhale silently to a mental count "1, 2, 3, 4" and then count *aloud* "1, 2, 3, 4, 5, 6, 7, 8" as you exhale. To make it more interesting, see how far you can count before you lose your breath and have to gasp for air. Imagine making such a desperate gasp in the middle of an important speech and you will realise why controlling your breathing is so important.
>   - Experiment with taking in a lungful of air in a quick but neat way, then releasing it slowly as you count aloud or recite a nursery rhyme, a poem or your favourite song lyrics. You can even call out the names of people in your class or favourite football team; anything will do, providing you practise and improve.

Voice box

The vocal chords are tiny and very delicate and can be damaged easily, for example, by shouting or screaming without having had a suitable warm-up.

Colds and flu can also affect the vocal chords and you may have heard instances where several members of a production cast have simultaneously fallen ill days before a show. You can help to avoid this by always warming up your voice using the exercises provided here.

Activity

As in the previous exercise, say some numbers aloud, but this time, place your fingertips on your throat. You may notice a slight vibration and movement on your larynx, which is sometimes referred to as the Adam's apple. This is a lump of cartilage at the front of your throat, easily noticed if you are male but more difficult to find if you are female. The larynx holds and protects the vocal chords (the two tiny flaps that vibrate to produce a wide range of sounds once the air from the lungs is pushed past them).

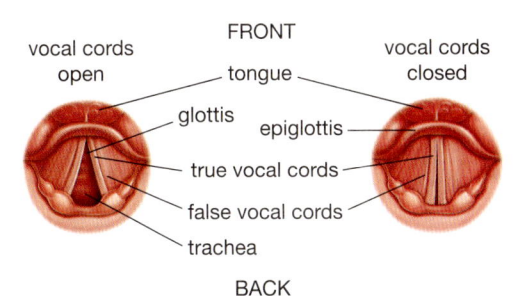

Figure 5.10 *Vocal chords*

Resonators

Imagine the strings on a guitar or violin as being the vocal chords; your throat and mouth are the resonators similar to the bodies of those instruments. If there is a lot of space in your neck and mouth your voice will be deeper. This is similar to the way that a cello produces notes deeper in tone than those of a violin. You should think of yourself as an extremely valuable instrument that you need to take good care of.

Try the following exercises to see how the shape of your mouth affects the sounds you make.

> **Useful tip**
>
> Keeping yourself free of germs by frequent and thorough hand washing is also important; it may seem silly, but doctors recommend it and it works, so why not do it?

Activities

1. By yourself, or standing in a circle with arms almost touching, try the following sequence:

   - Point your left hand out to the left, turn your head in the same direction and say "Hoo". Repeat this with your right hand pointing out to the right.
   - Point your left hand straight ahead and say "Hoe", then repeat with the right hand.
   - Point your left hand up and say "Hee", then repeat with the right hand.

   Repeat the first three steps, starting slowly and gradually increasing your speed. If you are working in a group, try keeping pace with those around you. The effect will slowly become like that of a steam engine: "Hoo, hoo, hoe, hoe, hee, hee". Make the sounds as round as you can and try dropping the pitch of your voice to make it deeper. See if you can be the last to stop as others become confused and drop out.

2. Experiment by yourself with sounds, just using changes of tongue position. Try having a conversation in "ape language" by just using vowel sounds. Aim to make it sound as convincing as you can.

> **Useful tip**
>
> The National Theatre website has a range of helpful videos on vocal warm ups, including exercises on resonance, articulation, and breathing.

Link

www.nationaltheatre.org.uk/backstage/voice

Developing your physical skills

Articulation

The word 'articulation' means 'joining together'. In voice work, it is used to describe how each sound is produced clearly, so that the words are heard and their meaning is understood. Always use the articulation exercises below as a warm-up before going on stage. If the audience can't follow what you are saying, then you are letting them, and yourself, down.

The three tools you use are parts of the instrument that is you: tongue, teeth and lips.

Tongue

Let's begin the articulation exercises with the tongue, as you have already considered it in regard to vowel sounds.

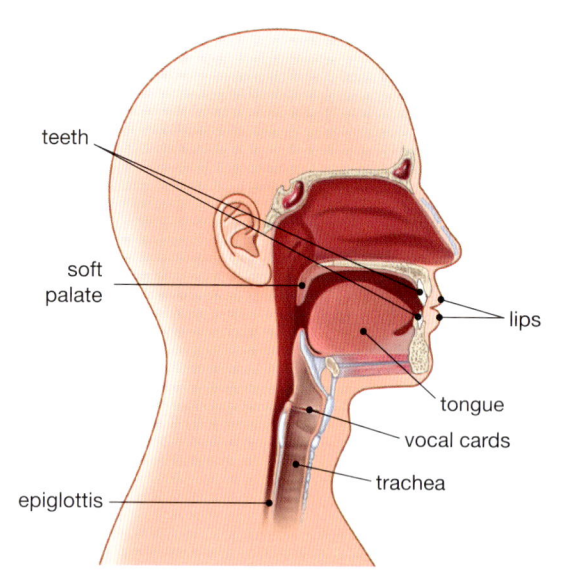

Figure 5.11 Your mouth is a resonator

Activities

1. Repeat the following steps four times to make a simple warm-up exercise.

   - One of the main consonant sounds modified by the tongue is the L sound. Try saying "Lu-lu-lu-la, Lu-lu-lu-la, Lu-lu-lu-la, Lu-lu-lu-la" over and over again, stressing the final "la". Then say "Lots of lovely little lively lemons." Practise and make perfect.
   - Now say "Gu-gu-gu-ga", repeated as above. Notice how the sound is made with the back of the tongue beating the back of your mouth.
   - Finally, say "Ku-ku-ku-ka" in the same rhythm as the others.

2. Say the phrase "Ragged rascal" in your normal speaking voice and then say it again, but this time rolling the R sound in an exaggerated way. If you find this difficult, practise rolls as if you are imitating a motorbike. Now say the phrase again, only this time with your tongue 'flapping' on each R. Once you have got the idea, say, loudly and confidently, "Round the rugged rocks the ragged rascals ran."

Teeth and tongue

Some sounds you make are created using your tongue and the alveolar ridge just behind your teeth. (Rather than "alveolar ridge", you can say "teeth" for ease.)

Useful tip

Invent your own tongue twisters and share them.

Activities

1. Say the words "Totally terrible" several times over and think about where you are placing your tongue. You will notice that it is just on the ridge behind your upper teeth.

2. Remember the exercises involving the L, G and K sounds and try them with the T sound, that is: "T-t-t-ta, t-t-t-ta." Make sure that each T sounds like a small explosion.

3. Say "D-d-d-dah, d-d-d-dah, d-d-d-dah" and, again, check your tongue position. Now, say the word "Today" several times and judge the differences between the two sounds (T and D).

The consonant sounds are very important, especially at the ends of words, so make sure you always sound them clearly. It may seem artificial on stage, but by the time the sound reaches the back of the hall, it will be natural. Trust your director, who will no doubt tell you if the ends of words are being lost. Remember that you will need to be aware of this, even if you have a sound system and microphones – these make your voice louder, but not necessarily clearer.

Be aware of losing important vowel sounds in the middle of words or phrases such as "Bottle" or "Got it". Instead of making the T sound, speakers often make a neutral sound in the throat called a **glottal stop**; this makes "Bottle" sound like "Bo'ul". Try saying "Lots of little milk bottles" and see if you are using glottal stops. If so, be careful in your next rehearsal, especially if it is for any play that requires the actors to speak in a Standard English accent.

Projection

Projection is the ability to throw your voice forward; it does not mean shouting. In fact, shouting is the worst thing you can do, as you can damage your vocal chords if you strain and try to force sounds from your throat.

If you place your hands on your abdomen, just below your rib cage, and breathe in deeply, you will feel the action of the diaphragm. This muscle controls the force of your breath and, therefore, its volume. The following exercises will help you to improve your projection.

> **Did you know?**
> A glottal stop is the sound of a consonant made by closing up the glottis, like when you say "ugh".

> **Useful tip**
> The following books are a good starting point for further individual research on the subject of voice work.
>
> Cicely Berry, *Voice and the Actor,* Virgin Books, 2000.
>
> Greta Colson, *Voice Production and Speech,* Longman, 1995.

Activities

1. Fill your lungs with air and recite part of a poem (try something you know well like a line from a production you are working on, or even a nursery rhyme). Try to make your delivery as loud as you can.

2. Now, whisper your recital, but focus on someone at the back of the room to make sure they can hear you. Imagine you have a volume control numbered 1 to 10. Your whisper is at '1'. Using your diaphragm to control your breathing, see if you can hit a '5', and then a '10'.

You can try this in two teams facing each other across the room or hall. Find a partner and take turns to speak comfortably and without strain. Show, by raising a number of fingers, how loudly you want your partner to speak, with '1' being the quietest. Watch and listen carefully and check for tension in the voice and body. If tension appears, go back to the relaxation.

Developing your physical skills

Unit 6 Preparing for a performance

To give yourself the best possible chance when you are performing your work for an audience, you should be as well-prepared as possible.

This section will help you to think about and prepare for all eventualities during the rehearsal period and up to the performance. Keeping records and updating them will save lots of time and enable you to get on with directing the piece or acting your role.

Objectives

In this unit you will learn to:

- be part of the creative team
- list all the equipment you have at your disposal
- find out how you will borrow or hire any other equipment
- keep constant communication with everyone in the group
- take responsibility for rehearsal schedules
- check all health and safety issues.

6.1 Putting your work together

A good performance depends on a few key principles, specifically:

- your practical understanding of repertoire
- your ability to devise and reflect on performance material
- your acting skills and ability to communicate effectively to an audience.

These are the areas you will be assessed on for the practical performance component of the Cambridge examination.

Figure 6.1 *Actors waiting backstage*

Preparing devised work for a performance

If you have a director, they will have the responsibility of leading the whole group in putting the work together and preparing for the performance. Otherwise the group must work together to make all the decisions and carry them out in order to achieve a successful performance.

As the work develops

Make sure that you are all aware of the contributions that everyone in the group will make. If you are working with a technical team, you must ensure that they are aware of all changes in the structure of the work and make time in rehearsals for experimenting with ideas. You need to make decisions about all staging elements as you go along so that you aren't in a rush at the end of your preparation time.

Here is a checklist of tasks to remember when preparing for your performance.

- Keep notes, diagrams and images from the very start.
- Organise your notes so that you can remind the group of any areas they are neglecting.
- Begin to make a **prompt copy** as soon as possible (see pages 116–117).
- Make sure that you have accurate copies of the script or text that is being used. Make any cuts clear.
- Consult with others so that you can tape out the ground plan as soon as possible.
- Make sure that there is sufficient space backstage for props tables and/or costume changes.
- Research sources for hiring costumes or lighting effects.
- Keep an up-to-date cast list (the performers may be playing more than one role).
- Find out if you have a budget to work with, and keep all receipts.
- Decide which **stage form** will be most appropriate.

Key terms

Prompt copy: a very detailed copy of the details of the performance with all cues (acting and technical) marked on it.

Stage form: the arrangement of the acting area and audience in your performance space.

Preparing for a performance

During rehearsals

The director should, as soon as possible, organise a **scenario** and make sure that everyone has a copy. However, you will need to be flexible as there may be changes. Provide rehearsal props and furniture for the sake of the actors. This will also help you, as you will be aware of what is on stage at any one time and will not forget to arrange how to **strike** it. If there are going to be costumes that may be difficult to wear, or complicated changes, suggest using practice clothes. At the end of each rehearsal, plan what you will do next time and, if you are working outside teaching time, draw up a rehearsal schedule. Give everyone a copy.

Preparing scripted work for a performance

If you are the director then you will be the person who will liaise with everyone to co-ordinate the group. Otherwise you may need to decide between you who will have responsibility for particular elements of the production.

However it works, you will also have a responsibility to the playwright. Some writers give very detailed stage directions of how the set must look and also detailed stage directions for the actors. It is not necessary to follow these to the letter but sometimes the details are essential.

For example, at the opening of Act 2 of *The Crucible*, the pot must be over the fire and the food ready for John Proctor to eat as he returns from planting the farm. It is important in showing the relationship between him and his wife.

It is interesting to consider the beginning of Jean Anouilh's version of *Antigone* in which he gives clear, detailed stage directions, including details of what the set could look like and the position of the actors on the stage.

> ### *Antigone*
>
> ANTIGONE, *her hands clasped round her knees, sits on the top step. The* THREE GUARDS *sit on the steps, in a small group, playing cards. The* CHORUS *stands on the top step.* EURYDICE *sits on the top step, just left of centre, knitting. The* NURSE *sits on the second step, left of* EURYDICE. ISMENE *stands in front of arch, left, facing* HAEMON, *who stands left of her.* CREON *sits in the chair at right end of the table, his arm over the shoulder of his* PAGE, *who sits on the stool beside his chair. The* MESSENGER *is leaning against the downstage portal of the right arch.*
>
> *The curtain rises slowly; then the* CHORUS *turns and moves downstage.*
>
> Extract from *Antigone*, by Jean Anouilh (translated by Lewis Galantiere)

Now compare it with this translation of Sophocles' original play, where the information given is much more condensed.

> *The scene is set outside the royal palace of Thebes.*
>
> *Enter Antigone and Ismene. They are both nervous and troubled. Antigone looks round to be sure they cannot be overheard before speaking.*
>
> Extract from *Antigone*, by Sophocles (translated by Don Taylor)

It is the job of the director to decide on what s/he would like the set to look like, as well as the concept behind the production and the style of the performance.

Key terms

Scenario: the summary or outline of the plot of the play and a list of the scenes in order and the characters which appear in them.

Striking (also known as bump-out): when the production is all over, the set, props and costumes are dismantled, packed and stored.

Did you know?

Upstage and downstage refer originally to the steeply raked stage of the Victorian theatre in the UK. So actors moving away from the audience were literally going upstage and those moving towards the audience coming downstage.

Making your prompt copy

- Write the details of each scene or use the pages of the script on one side and face this with a blank sheet.
- Write on the blank sheet the exits and entrances and where the actors are on the stage.
- Divide your acting area and use the abbreviations shown in Figure 6.2.
- The positions on stage are from the actors' point of view, i.e. 'right' means to the actor's right.

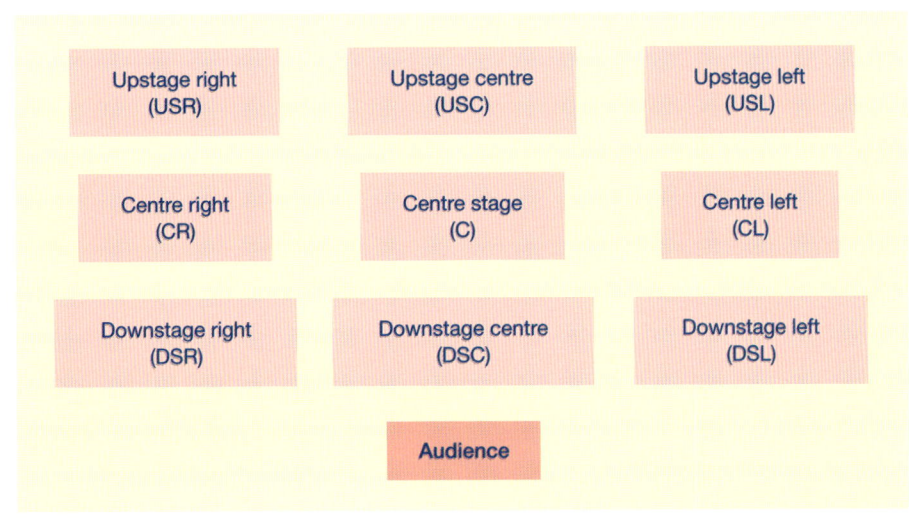

> **Did you know?**
> LX is short for electrics (that is, lighting) and SFX for effects (that is, sound).

Upstage right (USR)	Upstage centre (USC)	Upstage left (USL)
Centre right (CR)	Centre stage (C)	Centre left (CL)
Downstage right (DSR)	Downstage centre (DSC)	Downstage left (DSL)

Audience

Figure 6.2 These are the abbreviations for the different areas of the stage

> **Key terms**
>
> **Standby cue:** a warning to the operator to be ready for a change in lighting or sound.
>
> **Go cue:** an instruction to the operator to carry out a change in lighting or sound.

On the blank page, write all your cues and note in the script when they occur. Lighting cues are LX cues and sound cues are SFX. You should make a mark at the point where you need to give the sound and lighting operators their **standby** and **go cues**. Also mark in details of what these cues are and how long they are.

Once your group starts rehearsing the finalised work, find out when your performance date is and, by working backwards from that date, decide when your technical and dress rehearsal will be. Decide on these dates as soon as possible, to give everyone, including yourself, clear deadlines. Make sure that the space you need will be available.

Figure 6.3 The stage manager at work

On the next page is an example of what your prompt copy might look like. There is detail of the action and movement on stage, the set, the props, the lighting and sound cues, and when each happens.

3

THREE YEARS PASS AND MOTHER COURAGE, WITH PARTS OF A FINNISH REGIMENT, IS TAKEN PRISONER. HER DAUGHTER IS SAVED, HER WAGON LIKEWISE, BUT HER HONEST SON DIES.

A Camp

The regimental flag is flying from a pole. Afternoon. All sorts of wares hanging on the wagon. Mother Courage's clothes line is tied to the wagon at one end, to a cannon at the other. She and Kattrin are folding the washing on the cannon. At the same time she is bargaining with an ordnance officer over a bag of bullets. Swiss Cheese, in pay-master's uniform now, looks on. Yvette Pottier, a very good-looking young person, is sewing at a coloured hat, a glass of brandy before her. She is in stocking feet. Her red boots are near by.

THE OFFICER: I'm letting you have the bullets for two gilders. Dirt cheap. 'Cause I need the money. The Colonel's been drinking with the officers for three days and we've run out of liquor.

MOTHER COURAGE: They're army property. If they find 'em on me, I'll be courtmartialled. You sell your bullets, you bastards, and send your men out to fight with nothing to shoot with.

THE OFFICER: Oh, come on, uyou scratch my back, I'll scratch yours.

MOTHER COURAGE: I won't take army stuff. Not at *that* price.

THE OFFICER: You can resell 'em for five gilders, maybe eight, to the Ordnance Officer of the Fourth Regiment. All you have to do is to give him a receipt for twelve. He hasn't a bullet left.

MOTHER COURAGE: Why don't you do it yourself?

THE OFFICER: I don't trust him. We're friends.

MOTHER COURAGE *takes the bag*: Give it here. *To Kattrin*: Take it round the back and pay him a gilder and a half. *As the officer protests*: I said a gilder and a half! *Kattrin drags the bag away. The officer follows. Mother Courage speaks to Swiss Cheese*: Here's your underwear back, take care of it; it's October now, autumn may come at any time; I purposely don't say it must come, I've learnt from experience there's nothing that must come, not even the seasons. But your books *must* balance now you're the regimental paymaster. *Do* they balance?

SWISS CHEESE: Yes, Mother.

MOTHER COURAGE: Don't forget they made you paymaster because you're honest and so simple you'd never think of running off with the cash. Don't lose that underwear.

SWISS CHEESE: No, Mother. I'll put it under the mattress. *He starts to go.*

THE OFFICER: I'll go with you, paymaster.

MOTHER COURAGE: Don't teach him any hanky-panky.

Without a good-bye the officer leaves with Swiss Cheese.

YVETTE, *waving to him*: You might at least say good-bye!

MOTHER COURAGE *to Yvette*: I don't like that. *He's* no sort of company for my Swiss Cheese. But the war's not making a bad start. Before all the different countries get into it, four or five years'll have gone by like nothing. If I look ahead and make no mistakes, business will be good.

Figure 6.4 Sample of a prompt copy using the opening scene of Mother Courage, *by Bertolt Brecht*

Directions are not tied to specific moments in the text

On stage.

Wagon CS.

Box DSL

Cannon USR.

flag + washing

MC + K. SR

by cannon (washing)

Officer C.S (– bullets.)

S.C. DSL on box (bag)

Yvette USR – leaning on wagon.
(Hat + brandy)

M.C. cross to officer

M.C. confronts him

Takes bag

Turns back on him +

crosses SR to K + hand bag over

Kattrin exits USR + behind wagon

Officer crosses R + exits after K.

M C crosses to D.L. to S.C.
takes laundry

SC stands + takes laundry
starts to exit SR.

Officer enters SR from behind wagon

They exit S.R.

Yvette crosses DSC

MC crosses to her

As the flag is flown

LX21 stand by
SFX13 stand by
Go LX20 (X fade (5 secs)
SFX13 Go
(fade in to level 4)
+ play through

LX21 stand by

LX20 Go
(X fade 5 secs.)
(show general fad)
+ focus centre stage
10 secs

During the rehearsal process

When the work is at rehearsal stage, you will need to make decisions about set and costumes.

Tape the set out onto the floor of your performance space so that rehearsals can be organised and lighting can be fixed. Work out where you will be during the performances so that you can easily communicate with the sound and lighting operators. You will need to collect props and costumes, and work out where they will be stored. Keep these secure, as lots of time can be wasted searching for a missing item. It sounds mean, but do not lend your props to other groups. Keep records of where the props will come on and off stage. Check where costumes will be if a quick change is needed.

By now you should know if you will have to hire or borrow anything. Members of staff in your school or college are good sources, but it is a good idea to check out hire opportunities in your area and keep a list of phone numbers. You may have to go out "propping" – or shopping for props – so be sure of your budget.

Figure 6.5 Actors rehearsing from script

Key tasks for production management

- Make careful notes recording all changes the director makes, such as entrances and exits, and any props or costumes required by the actors.
- Keep a record of where the action on stage will take place.
- Start to collect props and furniture for rehearsals.
- Begin your prompt copy with a **ground plan**.
- Record all changes, especially cuts in the text.
- Draw up a **rehearsal schedule** and make sure everyone has a copy
- Note any potential hazards and means of resolving them.

Key terms

Ground plan: a scale outline of the set, drawn as if from above with indications of flats and furniture marked on it.

Rehearsal schedule: a list of times and places of rehearsals with the names of actors who are needed.

- Plan your **get-in**. Where will you store props, costume and set, and how will you get everything to your performance space?
- Plan your strike or **get-out**. Who will be responsible for removing everything from your performance space at the end of your performance?

The technical rehearsal

- Try to have this rehearsal a couple of days before the dress rehearsal, as this will give you some time to sort out any problems.
- Keep a list of contact numbers for your group in case anyone needs to get in touch with you.
- By now, your prompt copy should be complete and you need to have recorded all entrances and exits and where performers are on the stage throughout.
- You will also have pencilled in all lighting (LX) cues and sound (SFX) cues. You also need music and projection cues and anything else which is relevant.
- Decide where you will be during the performance so that you can cue technical operators and work out your cuing system.
- Go through the play from **cue to cue** and, when you are all satisfied with each one, note the details in your prompt copy in as much detail as possible.
- You may need to rehearse costume changes, complicated entrances and exits and any changes to the set.

The dress rehearsal

- Run this rehearsal exactly as a performance.
- If anything goes wrong, keep going and sort out the problems later. After all, if something goes wrong in the performance, you can't stop and start again.
- Make sure that the opening sequence is worked out so that lighting and sound work together.
- Check that every prop and costume is in place before you start.
- Check that all equipment is in working order.
- Make sure that the backstage area is tidy and safe, and make sure that you strike efficiently at the end.
- You will probably be sharing your performance space with other groups, so organise where you are going to keep all your props etc. so that they do not get in anyone else's way.
- You should leave your prompt copy in your performance space, and it should be so clear that anyone should be able to run the show from it.

If all goes according to plan, your performance(s) should run like a dream.

Key terms

Get-in: moving everything from storage and van onto the stage and preparing for the performance.
Get-out: removing everything from the stage to storage or van.
Cue to cue: go through the play to all moments when there is any technical change (to lighting, sound or set) and rehearse them.

Useful tips

- Keep in mind health and safety regulations
- Tape down loose cables.
- Keep notes of what you have done.

Remember

Warn everyone that this will be a long rehearsal and ask the performers especially to be patient.

Unit 7 Writing about your performances

Written examinations will normally be taken near the end of your drama course and will provide an opportunity for you to demonstrate your knowledge and understanding of practical and theoretical work you have undertaken based on a given text and stimuli.

Objectives

In this unit you will learn

- what skills and knowledge are tested in written examinations
- how to prepare yourself effectively

7.1 Written examinations

Drama is a practical activity but it is important to be able to show that you have understood what you were doing in your work and can reflect on it. You will need to be able to use the appropriate terminology you have learnt as an actor, director and designer.

Details of the written examination for the Cambridge syllabus are given on pages 4 and 5.

Thinking about the skills you have developed

Throughout this book, you have been advised to keep notes and records of the work you have done. You will also have a copy of the extract from a play that you have studied, and will have made copious notes on it as you worked through the scene from the point of view of actor, director and designer. You will not be able to take this into the examination room; there will be a clean copy for you to use instead. Don't worry – if you mark your personal copy with striking notes and memorable images, it may help fix the points you made when studying, and make them easier to remember when faced with a blank text.

Be ready to discuss clearly:

- the vision and intention behind your devised piece (the directorial concept)
- the ways in which you presented that concept to your audience and how and why you decided upon them
- the choice of performance space and its appropriateness to the directorial concept
- the process of development from ideas, workshop sessions and rehearsals through to performance
- more than one aspect of design used in your piece with details of how and why they were used (a question may ask for only one example, but it is best to be prepared for more)
- the effectiveness of the overall piece and of each of the contributory factors. Consider carefully what did and did not work and give reasons as well as examples.

> **Useful tip**
>
> If the question asks for more than one kind of response, for example, 'What?' and 'How?', make sure you give equal attention to each part. Weaker answers often concentrate on the first part, 'What?', and lose potential marks because there is insufficient evaluation.

Figure 7.1 The final bow

Effective preparation

- As soon as you are given the pre-release material, make notes on everything you do as you work on it. Involve yourself in discussion with your colleagues.
- Make sure your notes on each piece of work based on the play extract and the stimuli for devised work cover acting, direction and design.
- If you have not had an opportunity to play every role or consider design aspects in detail, then spend time sharing ideas and responses.
- Reflect on what others have said and go back to the text and decide whether you agree or disagree with the interpretations of others. Make sure you are armed with reasons and details to help you answer questions effectively.
- Be sure you know the reasons *why* decisions were made and can explain them clearly.
- Be sure you can explain clearly *how* work was carried out.
- Organise your notes according to the types of questions you will be asked.
- Look at and think about the points made in the next section.
- Keep calm and be confident – you have completed your practical work and you will be able to draw upon your experience and knowledge when answering questions.

Figure 7.2 An actress performing in a school performance of King Lear

Questions based on an extract from a play
From the perspective of an actor or a director

Remember that while an actor may focus on just one character, the director will have to be aware of all the characters and their interactions.

As an actor and a director, you will have had to interpret a character and find ways of communicating that character on stage. In doing so, you will need to have thought about the following:

- **Genre:** what is the genre of the play you are studying?
- **Style:** what is the most appropriate style of presentation for the piece?
- **Interpretation:** what is revealed about the characters from the words they are given? How does this fit in with what others say about them? Is anything implied about them in the stage directions by the playwright?
- **Characterisation:** how do you communicate the character's age, status and mannerisms?
- **Voice:** what accent, pitch, tone, pace, pause and emphasis will be most appropriate? When and how will you employ them? Why?
- **Movement:** what kinds of gesture, posture, gait, mannerisms and physical theatre skills could you use to make the character convincing to an audience?
- **Facial expression:** what changes of facial expression will best convey the character's changes of mood? When? How? Why?
- **Communication of relationships with others on stage:** when and how can you effectively use eye contact, physical contact and use of space to show what your character feels or thinks about other characters in the play?
- **Impact on your audience:** what kind of impact did you hope to have on your audience? What did you do to achieve this? To what degree were you successful?

From the perspective of a director and designer

The following three aspects are central to the realisation of any drama presentation:

- style
- genre
- period.

At an early stage of study and planning, you will need to have a very clear idea of where your piece fits into each of these aspects.

When planning the design and technical elements you will have to consider:

- **Costumes**: consider style, fabrics, textures and colours with details and reasons for choices. Think about ways in which costume conveys period, character and status.
- **Set**: think about the use of space and levels as well as the ways in which location can be represented, e.g. by flats, projection, materials, colours and textures. Consider how the set may be dressed in order to show historical period (past, present or future).
- **Make-up, masks and puppets**: consider these in terms of style, proportions, colours and appropriateness. What do they add to the overall effect?
- **Lighting**: consider use of colours, intensity, plot and special effects, and their effectiveness.
- **Sound**: is it live, recorded or a mix of both? **Diegetic** or non-diegetic (e.g. commentary or music)?
- **Properties**: how are they used and managed? How do they contribute to the effectiveness of the piece?

In all cases, while engaged in practical work you must be aware of health and safety factors such as the need for fireproofing (sets), awareness of skin allergies (make-up) and the need for electrical safety (lighting and sound).

Every part of any production needs to fit together and it is important to show that you have thought about the integration of individual elements into the overall design concepts. For example, the choice of sound effects and levels need to fit appropriately with the lighting, the colours of the set and costumes, and the moods evoked by the acting.

Remember that an actor will also be concerned with all of the points above, so it is important to communicate with the cast.

> **Remember**
> Always keep notes of strengths and weaknesses in the work in progress to help answer questions on evaluation as well as improve the overall quality of the piece.

> **Key term**
> **Diegetic:** sounds that belong to the world of the scene itself (such as footsteps or people talking), rather than effects that are later added (such as music or commentary).

Questions based on work devised in response to a stimulus

The points made in the sections above apply equally to devised work as well as a scripted piece. The big difference with devised work is that you will have had a part in creating the piece and therefore, instead of interpreting someone else's intentions, you have to make the decisions yourself. You will need to be very clear as to why you have made each choice. Be aware of how effective your decisions were in performance. To what extent did they communicate your intentions? Be honest, self-critical and able to support your points with evidence and details. As you work on your practical pieces, be prepared to listen to criticism, negative as well as positive, and keep notes.

In the examination room

- Always read the instructions on the front of an examination paper – they are there for your benefit.
- If it states 'write only in dark blue or black pen' then do so.
- Take note of the number of questions you must answer – if asked to answer *all* the questions in a section, make sure you do that. Words emphasised in **bold** are important details you must comply with.
- Don't rush into writing immediately – you are advised to spend some time reading the questions and the extract before you begin. Time reflecting and calming down is time well spent.
- The marks allocated for each question, e.g. [10], should give you an indication as to how long to spend on your answer. Manage your time carefully.

Figure 7.3 *Actress rehearsing Twelfth Night*

Useful tip

It may seem a difficult and daunting task to take in all the points made so far but remember, you will be tackling them all as you work on a range of practical pieces throughout your course. If you take each stage one step at a time and follow the advice given in this book, you will be in a position to tackle any question with confidence.

Remember

We learn best by taking risks and making mistakes. Don't be scared to make bold choices – if they don't work, try to recognise why they didn't.

Did you find that closing the curtain between each scene slowed the pace? Replacing it with a fade out and fade in of lighting would speed it up and make it flow more smoothly. Was the directorial concept made clear to your audience by your bold choice of costume or style, or were they just confused? Read some reviews of current theatre productions in your own country and see that even the professionals don't always get it absolutely right.

7.2 Practising exam-style questions

It is possible to obtain copies of Cambridge past papers and examiner reports, as well as the specimen pre-release material and the specimen paper, and mark schemes that go with them. These will help you to understand what will be required of you. They are available online at www.cie.org.uk.

As an indication of the kind of question that you might have to face, we have devised some practice questions. These are not officially approved by Cambridge but are the authors' view of the type of content you could expect. Remember that before examination you will have already spent time working on a given play extract from a practical point of view. You will also have devised and performed in response to one of the stimuli in the pre-release material.

Link
www.cie.org.uk

Activity

- Turn back to Figure 6.4 on pages 116–117 in Unit 6 'Preparing for a performance'.
- Read and study the extract from Bertolt Brecht's *Mother Courage* as if it were part of the pre-release material. (Note: there would, usually, be several more pages, but for the purposes of this exercise we have kept it short. Examples of the kind of stimulus you can expect are given later in this section.)
- Consider the extract from the point of view of director, actor and designer.
- When you think you are ready, try to answer the questions in the sample paper on the opposite page. (You will not be able to attempt Questions 7 and 8 until you have worked on a devised piece.)

Figure 7.4 The end of the performance

Section A

Answer **all** questions in this section

Questions 1–6 are based on the extract from *Mother Courage* by Bertolt Brecht, that you have studied.

1. Suggest the kind of costume Yvette Pettier and Swiss Cheese might wear and give a reason for your choice. *(2 marks)*

2. Identify **two** props that are used for dramatic effect and state what impact they have on the scene. *(2 marks)*

3. Suggest sound effects that might be appropriate to introduce the scene and establish its location. *(3 marks)*

4. Look at the dialogue from the opening speech by The Officer ("I'm letting you have the bullets for two guilders…") to line 17 ("I don't trust him. We're friends."). What aspects of The Officer and Mother Courage would you want to emphasise in performance? *(3 marks)*

5. As an actor playing the role of Mother Courage, how would you deliver the speech starting line 18 ("Give me the bag…") and ending with line 27 ("…Do they balance?")? *(5 marks)*

6. Summarise what you think would be the most important considerations in creating a set design for the extract. *(5 marks)*

Questions 7–8 are based on the piece of drama that you have devised from your chosen stimulus.

7. Choose one section from your devised piece that was particularly effective and say how you achieved this. *(5 marks)*

8. How did you structure your devised piece in order to communicate your directorial concept? *(5 marks)*

How well did you answer the questions? Did you manage your time carefully, giving more space and time to the questions that offered the most marks? There is a danger that because the questions appear to be simple and straightforward, you may be tempted to answer them at length, but remember you will need to divide your time across the whole paper according to the marks available.

Useful tips
- Time management is a key to exam success.
- Read the paper carefully.
- Note the marks available for each section.
- Note the number of questions and marks for each section.
- Allocate time to allow for reading, thinking about and writing your answer.
- Make sure you write the title of the stimulus you have used at the top of your answer so it is clear to the examiner.
- Practise timed answers to give you an idea of your own working speed.
- Do some calculations before you go into the examination room, then adjust them if necessary.

Section B

Answer **one** question from this section

Questions 9–11 are based on the extract from *Mother Courage* by Bertolt Brecht, that you have studied.

9   As an actor, what would you wish to communicate about the character Mother Courage as shown in the extract, and how would you achieve this in performance? *(25 marks)*

10 *Mother Courage* is a play set in a particular time and location. How could a set design for the extract help achieve this and how might other design elements be used effectively? *(25 marks)*

11 Describe how, as a director, you might wish to pace the performances in the extract from *Mother Courage*. Give clear details with reasons for your choices. *(25 marks)*

As you will see from the numbers in brackets, the questions in Section B carry more marks which means that you will need to spend more time and thought on these answers. First, read each question carefully and choose the one that you feel you can answer with most confidence. Think about what the question is asking carefully. For example, Question 9 requires you to:

- answer from the point of view of an *actor*, so you must draw mainly upon your knowledge of acting skills
- notice that there are two parts for you to answer: "What?" and "How?"
- focus chiefly on Mother Courage, but remember that her interactions with others on stage can be revealing
- consider *how* and *where* these skills are used in the communication of character – posture, voice, tone, gesture, gait.

Useful tip

Students who have practical working experience of the pre-release material are most likely to be able to give detailed and appropriate answers that gain high marks.

Remember

You should have some practical knowledge of the role you are writing about, having either played it or seen a colleague play it. Read through the script as if you are new to it and apply the skills described in Sections 2.3 'Creating a character' (pages 12–15) and 2.4 'Preparing your individual piece of work' (pages 16–21).

Section C requires a practical knowledge and experience of devising and performing a piece based on one of the stimuli offered in the pre-release material. For the purpose of this exercise, let us assume that these are:

- Stimulus 1: poem *Money* (*rant*) by Benjamin Zephaniah on page 30
- Stimulus 2: theme "First day blues" – for ideas on how to approach the theme of starting school, see pages 50–1 of section 3.12 'Devising using a specific audience'
- Stimulus 3: photograph of *Billy Liar* in Figure 3.17 on page 46.

You may wish to devise a short piece based on one of these stimuli before attempting one of the questions in the next section.

Though these are *not* official Cambridge questions, they are here as a guide to what you might expect to face in a written examination. Note that this section only requires *one* answer, so choose your question carefully based on your confidence in being able to focus on what the question is asking of you, and the amount of technical detail you are able to submit. Note the words in bold and obey them, for example, **either/or** in Question 14

Section C

Answer **one** question from this section

Questions 12–14 are based on the piece of drama that you have devised from your chosen stimulus.

12 What were the intentions behind your devised piece and how successful do you think you were in communicating these to your audience? *(25 marks)*

13 What drama skills did you use in presenting your character and how effective were these in performance? *(25 marks)*

14 Discuss how **either** costume **or** set contributed to the effectiveness of your devised piece. *(25 marks)*

obviously means you must make a choice – it sounds obvious, but it's easy to forget to read carefully when under examination pressure.

These longer-answer questions are worth up to 25 marks and are therefore likely to require more explanation and/or discussion than those worth 5 marks or less, so make sure you make sufficient reference to dramatic techniques, design elements and performance considerations. In your head you may have a very clear recollection of what you did and how and why you did it – make sure you write it down.

Avoid the temptation to present your answer in narrative form. Answers should focus on practical detail, not stories, so be particularly careful when answering on devised work. Demonstrate your knowledge and understanding of the processes you have followed.

A detailed account of the practical skills used and an evaluation of their effectiveness in performance is far more valuable than a fascinating and complicated plot.

Always support your answers with practical examples and detail – demonstrate your understanding of what you have learnt during your course.

If you are asked to comment on the effectiveness or success of your work, give an honest appraisal with specific examples. A missed cue, a misplaced gesture or a speech delivered with inappropriate emphasis may have been embarrassing for you at the time, but show that you are aware of such weaknesses as well as your strengths in your written answers. Demonstrate that you are fully aware of what you could change or develop in future performances in order to improve. The best work can always be improved – even this book!

Finally, enjoy your practical work, for that is what will help you make progress. We hope that this book has been of help to you and that you will continue to have a lifelong interest in drama. Good luck!

7.3 Preparing for questions about devised work in response to a stimulus

As you have seen in the previous sections, some examination questions focus on practical devised work you have completed in response to the pre-release material stimuli. Here is an example of one kind of stimulus you may be given:

Sample stimulus 1 – 'The good citizen'

What makes a good citizen?

You can adapt the work in this case study to suit your group size. The performance is linked to citizenship.

This project will enable you to explore a number of complex issues through movement and narrative form. You will need to explore:

- a wide range of ethnic issues and perspectives
- peer group pressure
- dangers to young people in today's society
- what young people perceive as dangerous.

The issue of 'citizenship' is universal in its appeal. Why do you think this is the case? What has happened in society to provoke this? These and other questions should form part of your planning. Figures 7.6 and 7.7 will also give you some starter activities to explore before you begin. You should then ask yourselves what are the intentions of your performance piece and who is your target audience?

Figure 7.5 Atticus Finch in To Kill a Mockingbird *is a good citizen*

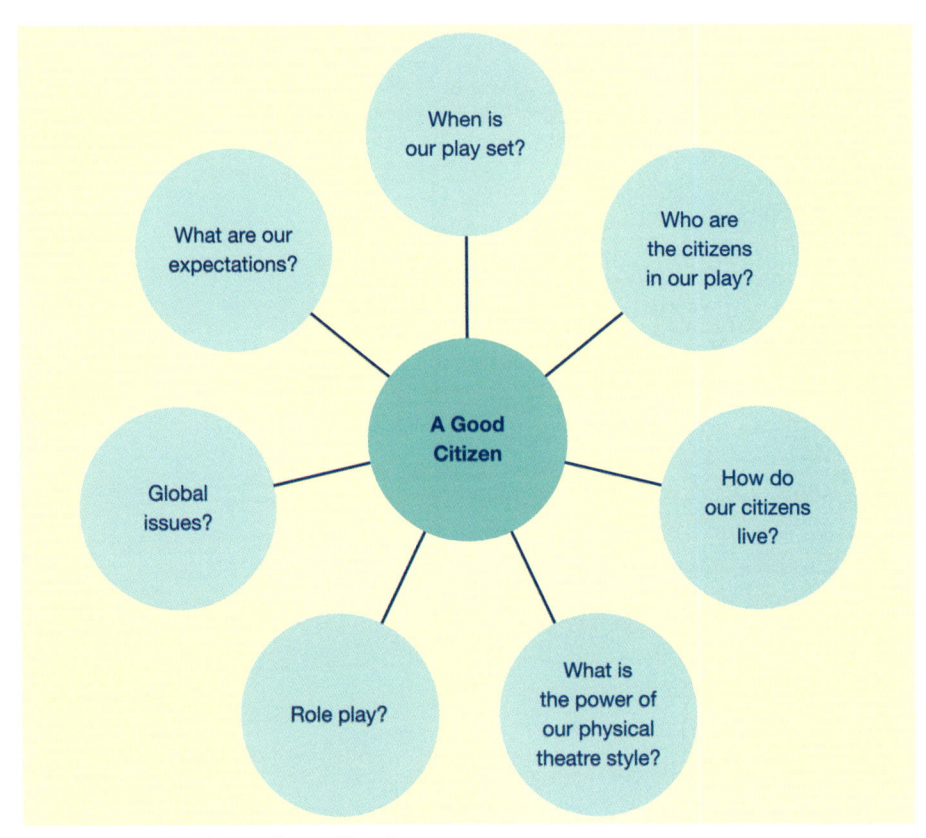

Figure 7.6 *Thinking about the theme*

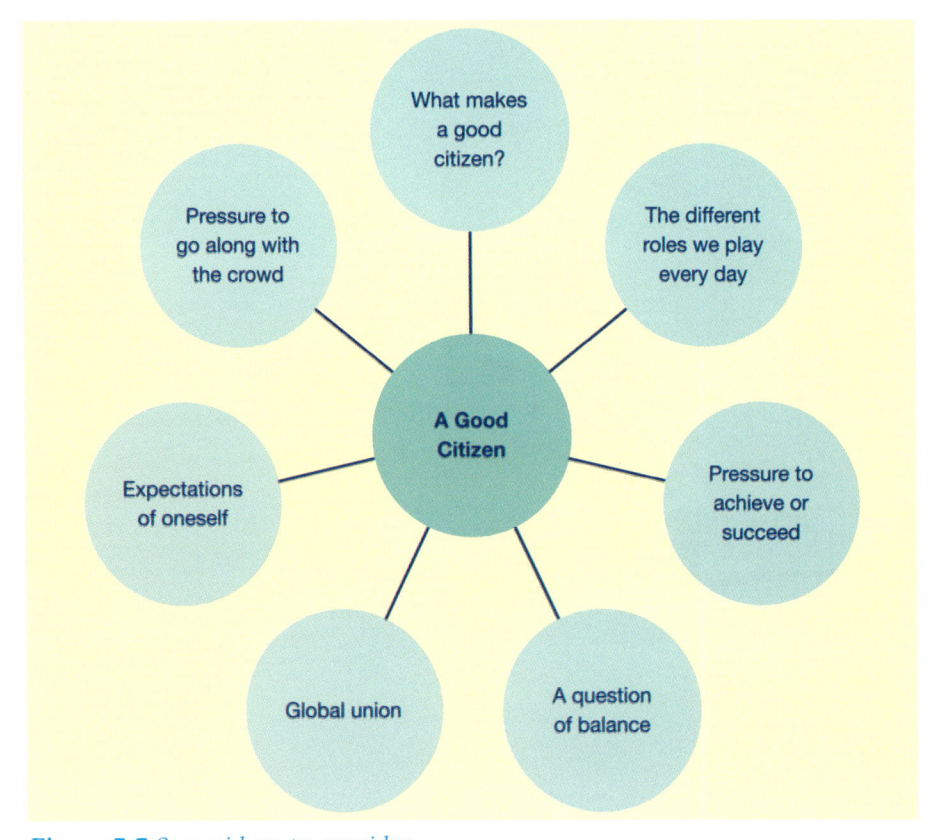

Diagram: Central circle "A Good Citizen" connected to:
- When is our play set?
- Who are the citizens in our play?
- How do our citizens live?
- What is the power of our physical theatre style?
- Role play?
- Global issues?
- What are our expectations?

Figure 7.7 *Some ideas to consider*

Diagram: Central circle "A Good Citizen" connected to:
- What makes a good citizen?
- The different roles we play every day
- Pressure to achieve or succeed
- A question of balance
- Global union
- Expectations of oneself
- Pressure to go along with the crowd

Useful tip

You should ask yourselves about your audience and how you will present your theme in performance terms.

Writing about your performances

Sample stimulus 2 – "Money"

Here is another example of the kind of stimulus material you may be asked to respond to in writing.

> ### Money in music
> In this case you could find a starting point in music, for example:
> - "Money, money, money" by Abba from the album *Arrival*
> - "Money makes the world go round" from the musical/film *Cabaret*
>
> or any other songs that you can immediately think of or gain access to that refer to the theme of money.
>
> Search for pieces or songs from different styles and music genres so that you discover a more diverse range of ideas.

Rather than just talk, get up and move to the music. What is being communicated in your moves? Write it down. Share your ideas and experiment with them. Listen to the others' views. Use the devising skills you have learnt previously to shape and present your ideas.

Remember to prepare sufficient material to allow you to answer a range of questions on your devised work.

Figure 7.8 The theme of money can be used as a stimulus

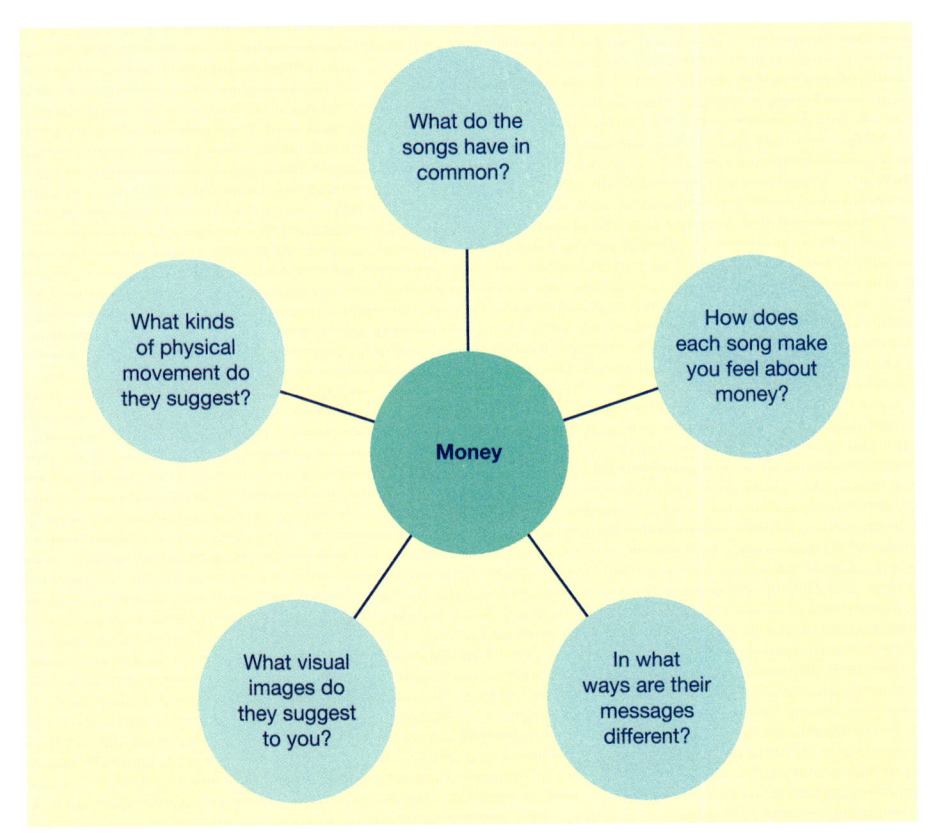

Figure 7.9 *Some ideas to discuss in your group*

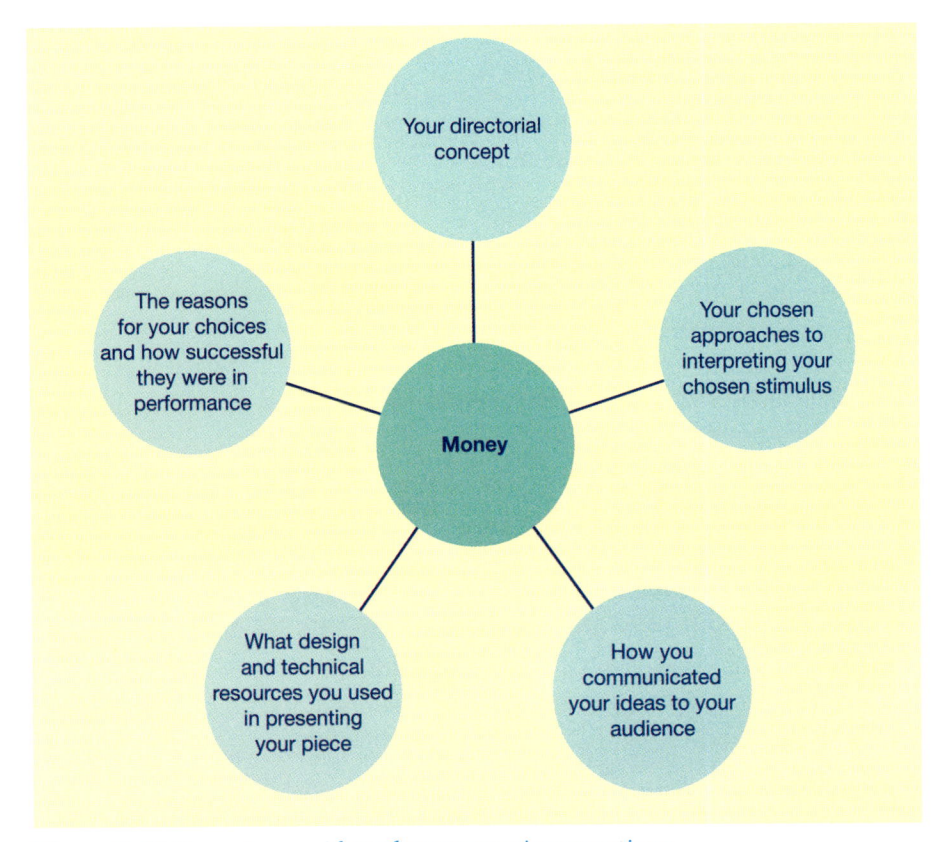

Central node: **Money**
- What do the songs have in common?
- How does each song make you feel about money?
- In what ways are their messages different?
- What visual images do they suggest to you?
- What kinds of physical movement do they suggest?

Figure 7.10 *Areas to consider when answering questions*

Central node: **Money**
- Your directorial concept
- Your chosen approaches to interpreting your chosen stimulus
- How you communicated your ideas to your audience
- What design and technical resources you used in presenting your piece
- The reasons for your choices and how successful they were in performance

Writing about your performances | 133

Glossary

A

Accessories: items of clothing such as hats, belts, ties or jewellery that add to the overall effect of a costume.

Angle: the direction from which the light comes onto the set.

Apron: an area of the stage that extends beyond the curtain line into the auditorium.

Articulation: the joining together of speech sounds to produce clarity of speaking.

Aside: a remark or comment that, unknown to others on stage, is directed at the audience.

Assemble: put together items of costume that you have found to make a complete outfit.

Audience participation: directly involving the audience in the production, for example, by asking them questions or giving individuals simple tasks.

B

Backdrop: a painted cloth (back cloth) hung at the back of the stage to set the scene or location of a play.

Balance: giving fair attention to other viewpoints so that the production is seen to be unbiased; this is very important when dealing with controversial topics.

Barn doors: generally four flaps (two opposing pairs) on the front of a lamp that can be opened or closed to adjust the width or shape of the beam of light.

Black light: use of ultraviolet lamps that cause specially treated surfaces to glow vividly when switched on; it is effective in puppetry and to give the effect of objects flying or moving by themselves on stage.

Blackout: a lighting cue for all lanterns to be switched off, plunging the stage into complete darkness.

Blocking: being told by the director where to stand, move or sit as you go through the first reading of the play; you can make notes of these moves in your script to help you to remember them in the next rehearsal.

C

Characterisation: the way in which an actor presents a character in a play.

Choreography: planning and design of dance and movement on stage.

Chorus: a group of actors who perform together and comment on the action of the play. In Ancient Greek theatre the Chorus sang, danced and declaimed between the main episodes and narrated the off-stage action. The Chorus in Elizabethan drama didn't sing or dance, but served a similar function, such as in Shakespeare's *Henry V*.

Colour filter: a coloured film placed in front of the lens of a lantern to change the colour of the light on stage (also known as a gel).

Colour wheel: a disc of coloured filters that rotates to produce a rainbow effect.

Composite set: a set on which it is possible to perform scenes in different locations or rooms in the same building without having to change the set all the time.

Conflict: an element of struggle, found in all drama; it may involve trying to resolve a problem or someone changing their life; it does not necessarily mean an argument.

Context: the background information surrounding a play, which helps us to understand the events and the characters.

Cross-fade: when one lighting state fades down as another one fades up, either instantly or over a period of time.

Cue sheet (lighting): a list of the lighting changes throughout the production (also known as a lighting plot).

Cue sheet (sound): a list of the sound changes throughout the production (also known as a sound plot).

Cue to cue: go through the play to all moments when there is any technical change (to lighting, sound or set) and rehearse them.

Culture: how the characters in the play live their lives.

Cyclorama: a curved cloth or a plain back wall of a stage on which lighting is used to create the illusion of distance or sky. The word is often shortened to "cyc".

D

Devising: planning a production and working out how it can be performed effectively.

Diaphragm: a large muscle at the bottom of the rib cage that assists in breathing.

Diegetic: sounds that belong to the world of the scene itself (such as footsteps or people talking), rather than effects that are later added (such as music or commentary).

Dimmer pack: a number of dimmers used to control the lighting intensity on a set.

Direct address: an actor speaks directly to the audience.

Director: the person who tells an actor how and when to do something on stage.

Directorial concept: the basic ideas that underpin your production and that you want to communicate to the audience.

Documentary: putting factual information across to the audience.

Dress rehearsal: a full rehearsal (as of a play) in costume and with stage properties shortly before the first performance.

Dynamics: variations or contrasts in the energy, speed and movement you create in your performance.

E

Effects projector: a device used to project an image from a rotating glass disc to give the effect of, for example, clouds, flames or rain.

Ensemble: a group of people working together; everyone makes an equal contribution and there is no 'starring role'.

Expressionistic: drama that tries to show emotions rather than reality. The term also applies to other art forms and is often contrasted with "naturalism" or "realism".

F

Farce: a fast-moving light comedy often involving ridiculous characters and improbable situations.

Flashback: when the narrative in a drama switches from the current time (current from the point of view of the characters) to an incident from the past, perhaps as a memory or a dream by one of the characters.

Flat: a light wooden frame covered in scenic canvas, plywood or hardboard which can be painted to suit your work.

Focus: concentrate the lights onto a specific area of the set.

Forum theatre: an interactive form of theatre developed by Brazilian director Augusto Boal; the audience stop the play to suggest different solutions to a problem that a main character is experiencing.

Fourth wall: the invisible wall between the audience and the actors on stage, commonly associated with naturalistic theatre.

Fresnel: (pronounced fre-nel) the most common type of lantern used on stage, it has a textured lens and produces a very even light that is soft at the edges and tends to project a soft shadow.

G

Gait: the manner in which a character walks and moves around the stage.

Gauze: also known as a scrim, gauze is a coarse-weave fabric which appears transparent when the scene behind it is lit; sharkstooth is the most opaque.

Gel: a coloured film placed in front of a lantern to change the colour of the light on stage (also known as a colour filter).

Genre: a category of art, music or literature; drama is a literary genre. Drama is further divided into types of production such as tragedy, comedy, farce and melodrama.

Get-in: moving everything from storage and van onto the stage and preparing for the performance.

Get-out: removing everything from the stage to storage or van.

Gobo: a small, perforated metal sheet placed between the lamp and the lens of a spotlight to project an image on to the stage.

Go cue: an instruction to the operator to carry out a change in lighting or sound.

Ground plan: a scale outline of the set drawn as if from above with indications of flats and furniture marked on it.

H

Hot-seating: the technique of an actor staying in role while answering questions from the audience about the character's thoughts and feelings; the actor can involve the audience by asking them for advice.

House lights: auditorium lights in a theatre that illuminate the audience before and after the performance.

I

Illusion: anything that deceives the senses by appearing to be something which it is not.

Improvisation: when actors invent or make up a script spontaneously. In devised theatre, this is a useful method to develop characters and situations in order to create a play. In IGCSE Drama, improvisation refers to any unscripted work.

Inflection: changes in vocal pitch, tone or volume in order to alter meaning or intent.

In role: appearing convincingly and consistently as a character different from one's self.

L

Lanterns: lights used to illuminate a set.

Lighting board: control desk for lighting.

Luminescent: a word that describes an item that glows in the dark when ultraviolet light is switched on.

M

Makes: items that cannot be found or borrowed and must therefore be made.

Mannerism: a distinctive and individual way of moving or speaking.

Melodrama: a 19th century genre of theatre in which music was used to intensify the usually sentimental or sensational story in which Good always overcame Evil. It is a style characterised by stock characters and exaggerated emotion and gesture.

Mime: using clear gestures and movements but no words to convey a character's personality, emotions and a narrative.

Monologue: when a character on stage speaks alone, sometimes directly to the audience.

Motivation: the reason why a character does something or behaves in a certain way.

Mummers' play: a traditional folk play performed usually at Christmas and involving characters such as St George.

N

Narrative: the story that your performance wants to tell the audience.

Naturalistic: attempts to faithfully represent real life on stage.

Newspanel: written messages flashing across a screen during the play, perhaps giving facts that the audience would find difficult to take in if they couldn't see them.

P

Pace: the speed and rhythm of your speech and how you pick up cues from others.

Pace-egg play: similar to a mummers' play but performed in the North of England, often on Good Friday.

Parcan: a lantern that is used to provide strong dramatic keylight, backlight or effects such as beams of light in smoke.

Parody: a satirical version of a more serious work with a view to mocking or undermining the original piece.

Patching: connecting stage lights (lanterns) to dimmer controls for fading in and out.

Period: the time period in which a play is set.

Physicality: a character's unique physical features (presence, gestures, posture etc.) which the actor will need to convey.

Posture: the ways in which one's body is held in order to communicate character, mood or emotion.

Profile: a focusable lantern with an ellipsoidal lens which enables a sharp beam of light to be projected. Also known as an ellipsoidal reflector spotlight (ERS).

Prompt: to remind actors of a line forgotten during a performance. Usually the responsibility of the stage management and conventionally located stage left – hence "prompt corner" and "prompt side".

Prompt copy: a very detailed copy of the details of the performance with all cues (acting and technical) marked on it.

Props: any object on stage that is not costume or set. Short for "properties".

Proscenium: an arch that separates the stage from the auditorium, common in traditional theatres.

Prosthetics: artificial body parts; an example in stage make-up could be nose-putty, moulded to change the shape of a nose then coloured by make-up.

Protagonist: the leading character in a drama – from the Greek word *protagonistes* meaning "the chief or first actor".

Proxemics: describes how close or far away actors are from each other on stage and how this is used to convey dramatic meaning.

R

Realism: a 19th century attempt at presenting real life on stage through recognisably "real" settings, costume and dialogue. It was to some extent a reaction to the exaggeration and heightened emotions of melodrama.

Register: the tone or formality of language used to convey a particular character or setting.

Rehearsal schedule: a list of times and places of rehearsals with the names of actors who are needed.

Rig: hang the lanterns in the correct positions.

Rostrum (plural rostra): a portable platform which you can use to create interesting levels.

S

Safe load: the maximum weight that should be put onto a lifting device or suspension point.

Satire: a play that uses humour, irony or ridicule to criticise and expose people's foolishness.

Scenario: the summary or outline of the plot of the play and a list of the scenes in order and the characters which appear in them.

Set dressing: including items on the set that add to the overall effect. For example, Figure 4.39 has cushions, plants and pictures.

Setting: putting props onto a set ready for a performance and taking them off again.

Setting line: an imaginary line marking the furthest point downstage that scenery may be set, usually just above the tabs (curtain) line. Marked on stage plans it offers a reference point for placing flats, scenery and other items on stage.

Sight line: what your audience can see on the stage; sit at the extreme ends of the front row to work it out.

Society: who the characters in the play are.

Soliloquy: a dramatic device by which the inner thoughts and feelings of a character are revealed by an actor speaking aloud as if to him/herself.

Sound plot: a list of sound cues and levels in running order.

Stage form: the arrangement of the acting area and audience in your performance space.

Stage presence: an ability to command the attention of an audience by one's manner or physical appearance on stage.

Standby cue: a warning to the operator to be ready for a change in lighting or sound.

Status: the relative position or standing of someone in a group to others, for example, the ship's captain has a higher status than the surgeon but lower status when ill.

Stimulus (plural stimuli): the starting-point for a devised work; the idea, image or object that sparks off your work.

Stock character: a farcical character in theatre, who reoccurs in a dramatic tradition, and is therefore quickly recognised by the audience and needs no introduction.

Striking (also known as bumping-out): when the production is all over, the set, props and costumes are dismantled, packed and stored.

Structure: the organisation of work in terms of its starting point, its setting and narrative line.

Style: the way in which the production is presented and performed, such as in a naturalistic fashion or physically.

Symbolic: a sign or object that represents or typifies something else; for example, a helmet and an axe might be symbolic of a fire station.

T

Tableau: the character/s, clearly showing their role, appear as a still image.

Target audience: the specific audience (defined, for example, by age or interest) for which a production is devised.

Theatrical conventions: generally accepted rules or ways that govern how a particular genre or period play is performed.

Theme: the central subject matter or idea supporting a dramatic work.

Tone: using your voice to express what you are feeling.

Transition: a change between scenes or sections of your work.

U

Upstage: (verb) to steal the attention from other actors, originally by standing physically upstage of them in a more dominant position. Not recommended unless specifically required by one's director.

V

Verbatim theatre: the words of a real person are recorded as part of the research for a piece of work. The words used – and every pause or "er" – are then spoken by the actor in exactly the same way in the performed piece.

W

Western: a genre in literature, drama and film which tells stories set in the American West during the late 1800s. Westerns often feature American Indians, the original inhabitants of the land, and cowboys, the new settlers.

Wings: areas at either side of the stage that are unseen by the audience.

Index

A

accessories 66
acting 9
  getting started 9
apron stages 88
articulation 110
  projection 111
  teeth and tongue 110–11
  tongue 110
assembling costumes 67
audiences 50
  audience participation 56
  consider involving your audience in the action 52
  involve your audience's sympathy 52
  plan your audience management 54
  research your audience 50
  research your materials 50–1
  research your theme 50

B

balance 51
barn doors 71
beginnings 57
blocking 10
breathing 107–8
bump-out 114

C

characters 12, 40
  building a physical picture 40
  building a vocal picture 41
  characterisation 36, 53
  communicating your character to an audience 44–5
  creating the role of Abigail Williams 14–15
  creating the role of John Proctor 13
  developing a character for devised work 42–3
  stock characters 99
colour filters 69
composite sets 24, 90
conflict 17, 44
context 16
costume design 61
  as the work develops 64
  costume design for scripted work 63
  during rehearsals 64–5
  effective costume design 66
  experiment through sketches 66
  generating costume ideas 61
  make or assemble 67
  planning your work 62
  research the character and consider materials 66
  research the period 66
cross-fades 24, 31
cues 115
  cue sheets 71, 94, 97
  cue to cue 49, 119
culture 9

D

design 7, 60
  costume design 61–7
  lighting 68–71
  make-up 72–7
  masks in performance 78–81
  properties 82–3
  puppets 84–5
  set design 86–93
  sound 94–7
devising 22–3, 25, 59
  as the work develops 113
  bringing your show together 56–7
  developing a character for devised work 42–3
  devising from a character 44–5
  devising using a specific audience 50–1
  devising work from a theme 26–9
  devising work using a character 40–1
  devising work using a photograph 46–7
  devising work using a poem 30–5
  devising work using improvisation 38–9
  during rehearsals 114
  getting started 23
  lighting 69
  make-up 74
  making your play 48–9
  performance skills 36–7
  performance styles 58–9
  planning the production 52–3
  preparing for the show 54–5, 113
  questions about devised work in response to a stimulus 130–3
  resources and skills 24–5
  set design 90–1
diaphragm 107
diegetic sounds 94, 124
direct address 19
directors 11
  directorial concepts 35, 50
documentary 27, 58
drama 2–3
  learning new skills 3
  working with other people 3
dress rehearsals 49, 119
dynamics 102

E

effects projectors 69
end-of-the-pier shows 62
end-on staging 88
ends 57
ensemble 23
expressionistic styles 58

F

facial expressions 105
flashbacks 26, 48
flats 86
forum theatre 52
fresnels 68

G

gait 21
gauze 24
gels 28, 70
genres 9, 58–9
gestures 104

get-ins 119
get-outs 119
glottal stops 111
go cues 115
gobos 28, 71
Greece, ancient 2
ground plans 87, 118

H

hot-seating 13, 42
   experiment with hot-seating 53

I

IGCSE assessment 4, 6–7
   coursework 5
   good practice during your course 5
   plan ahead 5
   written examination 4, 120–1
illusion 76
improvisation 38
   experiment through improvisation 56–7
   getting started 39
in role 52
individual work 16–18
   how will you move? 21
   how will you speak? 21
   importance of pace 18–19
   using your space 19–21
   where will you be on stage? 21

J

jumping 103

L

lanterns 70
lighting 68
   as the work develops 71
   cue sheets 71
   devised work 69
   during rehearsals 71
   focus 70
   lighting boards 70
   practical lighting 70
   scripted work 69

M

make-up 72
   as the work develops 75
   character make-up 72
   designing make-up for a devised work 74
   during rehearsals 73, 75
   fantasy make-up 72
   make-up design for The Crucible 74
   research different materials and products 72–3
   straight make-up 72
   work as part of the group 76–7
makes 83
masks 78
   as your work develops 81
   audibility 78
   characterisation 79
   doubling 78
   performance style 80
   visibility 78
materials 50–1
   shape your material 57
melodrama 59
mime 40
monologues 16, 44
motivation 53
movement 102
   gesture 104
   jumping 103
   stepping 102
   travelling 102–3
   turning 103
music 105

N

naturalistic styles 58
newspanels 27
non-diegetic sounds 94

O

organising your show 54

P

pace 41, 57
   importance of pace 18–19
pantomime 58
parcans 68
performance skills 36, 58
   characterisation 36
   rehearsing and performing 49
   use of movement 36–7
   use of voice 36
performance styles 58–9
period 9
photographs 46–7
   using photographs 47
physical skills 7, 98
   characterisation 40
   facial expressions 105
   introduction to voice work 106–11
   movement 102–4
   planning a performance 101, 102
   styles of physical theatre 99–100
   using physical skills in performance 99–100
   using your skills in developing a performance 102–5
physicality 40
plays 10–11, 48
   rehearsing and performing 49
   structuring your play 48–9
poetry 30–5
practical work 7, 22
   bringing your show together 56–7
   developing a character for devised work 42–3
   devising from a character 44–5
   devising using a specific audience 50–1
   devising work from a theme 26–9
   devising work using a photograph 46–7
   devising work using a poem 30–5
   improvisation 38–9
   making your play 48–9
   performance skills 36–7
   performance styles 58–9
   planning the production 52–3
   preparing for the show 54–5
   resources 24
   skills 25
   working from a character 40–1
preparing for a performance 7, 112–13
   dress rehearsal 119
   during the rehearsal process

118–19
  making your prompt copy 115–17
  preparing devised work 113–14
  preparing scripted work 114
  technical rehearsal 119
profiles 68
projection 111
prompt copies 113
  making your prompt copy 115
  sample of a prompt copy 116–17
properties (props) 82
  finding props 83
  making props 83
  personal props 83
  research widely 82
proscenium arches 87
prosthetics 76
proxemics 37
puppets 84–5

Q

questions based on an extract from a play
  from the perspective of a director and designer 124
  from the perspective of an actor or director 122–23
questions based on work devised in response to a stimulus 125, 130–3

R

register 49
rehearsals 49
  costume design 64–5
  devised work 114
  during the rehearsal process 118
  key tasks for production management 118–19
  lighting 71
  make-up 73, 75
  rehearsal schedules 118
  sound 94–5
research 50–1
  costume design 66
  make-up 72–3
  props 82
  turn your research into a production 56

resonators 109, 110
rigs 70
rostrum (rostra) 28, 86

S

scenarios 48, 102, 114
scrim 24
scripted work 7, 8
  acting 9
  costume design 63
  creating a character 12–15
  lighting 69
  looking at a play 10–11
  preparing for a performance 114
  preparing your individual piece of work 16–21
  set design 88
  sound design 95
set design 86
  be practical 92
  creating your own set design sheet 89
  examples of open stages 88
  fit in with other design elements 92–3
  keep to the concept 93
  set dressing 90
  using your space 87
  working on a script 88
  working on set design for a devised piece 90–1
setting lines 91
sight lines 91
society 9
soliloquys 17
sound 94, 105
  as the work develops 94
  cue sheets 94, 97
  during rehearsals 94–5
  managing sound for the performance 96
  recording and editing sound effects 96
  role of sound designer 96
  sound design for scripted work 95
  sound plots 95
space 19–21
  choose your performance space to fit the show 55

set design 87
stages 88
  stage form 102, 113
standby cues 115
status 39, 55
stepping 102
stimulus (stimuli) 22, 30
stock characters 99
striking 114
structure 29, 31, 39
  structuring your play 48–9
styles 16
  experiment with style 58
  genres 58–9

T

tableau 21
target audiences 50
technical rehearsals 119
technical work 7, 60
  costume design 61–7
  lighting 68–71
  make-up 72–7
  masks in performance 78–81
  properties 82–3
  puppets 84–5
  set design 86–93
  sound 94–7
theatre 2
theatre-in-the-round stages 88
theatrical conventions 58
thrust stages 88
tone 41
tongue twisters 110–11
transitions 31
travelling 102–3
turning 103

V

verbatim theatre 23
voice work 32–5
  articulation 110–11
  breathing 107–8
  resonators 109, 110
  vocal chords 108–9
  voice box 108–9
  warm-up exercises 106–7

W

warm-up exercises 106–7
Westerns 37
working as a group 52
  make-up 76–7
working from a poem 30
  group performance 35
  poetry and voice work 32–5
  structuring your work 31
working from a theme 26–27
  'despair' 28
  examples of text 29
  some points to consider 28
  structuring your work 29
writing about your performance 7
  effective preparation 122
  in the examination room 125
  practising exam-style questions 126–9
  preparing for questions about devised work in response to a stimulus 130–3
  questions based on an extract from a play 122–4
  questions based on work devised in response to a stimulus 125
  thinking about the skills you have developed 121
  written examinations 4, 120–1
written examinations 4, 120–1
  in the examination room 125
  practising exam-style questions 126–9